A
HOUSE
IN
SPACE

OTHER BOOKS BY HENRY S. F. COOPER, JR.

Apollo on the Moon
Moon Rocks
13: The Flight That Failed

A
HOUSE
IN
SPACE

Henry S. F. Cooper, Jr.

HOLT, RINEHART AND WINSTON
NEW YORK

Library of Congress Cataloging in Publication Data

Cooper, Henry S. F.
　A house in space.

　1. Skylab Project.　I. Title.
TL789.8.U6S5547　　629.44′5　　75-42991
ISBN　0-03-016686-1

Most of this book originally appeared in slightly
different form in *The New Yorker*.

PRINTED IN THE UNITED STATES OF AMERICA

10 9 8 7 6 5 4 3 2 1

All photographs in this book are
courtesy NASA.

To
Mary, Lizzie,
and especially
HANNAH,
who didn't get one before

A
HOUSE
IN
SPACE

*Oh, baby, when I was a little kid, I never
dreamed anything like this even could happen. Nope, I never dreamed it could happen
to anybody, let alone me.*
—EDWARD G. GIBSON, SCIENCE PILOT,
THIRD SKYLAB MISSION

ONCE EVERY FIVE DAYS, Skylab, the abandoned space station, orbits overhead. If its pass occurs just before dawn
or just after sunset, the empty hulk appears as a bright
star speeding purposefully across the sky. When the third
and final crew of astronauts to live there departed on
February 8, 1974, they closed down the space station in
such a way that it could easily be put back in commission:
When they deactivated the ventilating system, they replaced with new units the charcoal canisters that purified
the air; and when they drained the water from the pipes,
they left the valves open so that the plumbing would be
clean. They even left a bag near the entrance hatch containing food, film, and a number of other items that any
future visitors might bring back to earth with them, for
scientists are interested in the long-term effects of space
on many materials. Any future visitors to Skylab won't
stay long, though, for even before the third Skylab crew
went home some of the space station's systems, in particular the one for guidance and navigation, had begun
to fail. Indeed, as the third crew of Skylab astronauts
pulled away from the space station, one of them said
later, they felt it was the end, for they knew that no one
would ever again *live* aboard Skylab, floating weightless
through its rooms, wrestling with experiments, and just
peering out the window for hours at a time, as they had
done since their arrival on November 16, 1973.

On that day, the space station—gold, white, and silver

—was hard to make out against the swirling white clouds of the earth 269 miles below. Skylab resembled a huge, squat helicopter with a tower overhead surmounted by what looked like a big four-bladed rotor, but which was in fact an array of solar panels for generating electricity. It was somewhat battered now. After six months in space, the white paint had browned slightly, and some of the gold had baked and blackened. It was minus one of a pair of stubby, winglike solar panels that had broken off shortly after it was launched from Cape Kennedy on May 14, 1973; in the mishap, insulation protecting it from the sun had been shredded from its surface. The first crew, which had arrived twelve days later, on May 26, 1973, had erected a protective parasol over the space station; and the second crew, which had arrived on July 28, had spread a huge awning over that. Consequently, with its awning and its parasol, its pinwheel of solar panels overhead and its single remaining wing, Skylab now looked less like a space station than like Uncle Wiggily's airship.

Indeed, some aerospace experts said that strictly speaking Skylab was not a space station, which is usually defined as a structure that, like a house on earth, can be resupplied indefinitely with such consumable items as fuel, water, and food. There were no fittings on Skylab for refueling it or for renewing its supplies of nitrogen, oxygen, and water; after an effective life of about nine months, it would have to be abandoned. Nonetheless, almost everyone at the National Aeronautics and Space Administration *thought* of Skylab as a space station, and the people there pointed out that it was, after all, resupplied with some items, such as film, experiments, and even astronauts. If it wasn't a space station, they said, it was hard to say what else it *was*. At the very least, Skylab was a prototype space station. It had been knocked together from old Apollo parts; indeed, NASA might have built a more advanced version if it had planned it from scratch instead of setting out to find a use for surplus Apollo hardware, which was how the project started. Otherwise, though, there might have been no space station at all. The astronauts sometimes called it "the cluster," and sometimes "the can," and both terms were correct, for Skylab was a cluster of cans. By far the biggest can, called the workshop, was a converted Saturn rocket booster of the sort that used to power Apollo astronauts out of earth orbit toward the moon; it was 48 feet long and 21½ feet in diameter. At its forward end was a shorter, thinner can, roughly the proportions of an old Apollo spacecraft. Inside this, up against the booster, was the airlock module, from which astronauts ventured on space walks; and further out was the multiple-docking adapter, at the tip of which the third crew landed. Skylab was certainly big enough to qualify as a space station by any definition; when the command-and-service module had docked and become part of the cluster, it had the volume of a three-bedroom house, weighed (had it still been on earth) almost a hundred tons, and was as tall as a twelve-story building. It was the biggest structure any astronauts had ever seen in space.

The three astronauts of the third crew floated head first out of the conical command-and-service module; through the docking adapter, a white tunnel of a room 17 feet long; through a hatch into the smaller airlock module; through another hatch into a tunnel that passed through a ring of gas tanks and other ventilating equipment; and on into the cavernous workshop, which filled what would have been a huge fuel tank inside the booster. (A smaller tank to the rear was used for garbage.) The workshop tank was divided into an upper and a lower deck, the upper one being by far the bigger; the lower, which they floated into last of all, was only a little higher than a man; and it was here, the furthest inside Skylab,

"With its awning and parasol, its pinwheel of solar panels overhead and its single remaining wing, Skylab now looked less like a space station than like Uncle Wiggily's airship." From above, and docking (*left*). From below (*right*).

that the astronauts' bedrooms and living quarters were. The long tunnels, ending in the huge chamber, made Skylab look comfortably womblike, as though NASA had had some sort of collective unconscious; indeed, while the astronauts were in the workshop, they floated about weightless and even adopted slightly crouching, almost fetal positions; and when they emerged many months later to go home, they were different physically and, some astronauts would later claim, mentally as well.

From the standpoint of time spent in space, as well as from the standpoint of the large size of the space station, both of which imposed special sets of circumstances, Skylab extended man's experience in a new way. The difference between it and earlier missions was between going on a quick trip through space in a vehicle the size of a car, and moving there to stay awhile in a house with all its rooms and corridors. The Skylab astronauts were the first to *live* in weightlessness. Like many long visits, the ones aboard Skylab were almost a state of mind. It was not suspenseful, like the expeditions to the moon, but a steady, continuous experience, like life anywhere.

On a normal day during the third mission, the three astronauts were awakened by a buzzer at six o'clock in the morning, Central Standard Time, the time at the Johnson Spacecraft Center outside Houston, from which the flight was controlled. They felt refreshed; they logged much more sleep than the Apollo astronauts, who had been under more strain and had not had their own individual bedrooms. Though the Skylab astronauts slept an hour or so less than they did on earth, they may not have needed as much sleep in space, where existence was less arduous; their pulses averaged about twenty beats lower, and while they slept their hearts contracted only thirty times a minute. Before they could get out of their beds, which were sort of sleeping bags hung vertically against the walls of their bedrooms, they had to unzip a light cloth that had kept them from floating off during the night. Then they soared upward out of bed as if by magic—as though they were genii escaping from bottles. They moved about without using their feet; and even if they made their feet go through the motions of walking, and then stopped these motions, they kept right on going until they hit something. If they thrust out an arm in one direction, they moved back in the other. They could not have experienced these phenomena as readily in the smaller craft used by earlier astronauts; by its size, Skylab added a dimension to weightlessness. The astronauts, and everything they handled, moved as though they were underwater, in a sort of dreamlike, disembodied way— or as though they were in a magical place. They *were,* of course, if by magic is meant a suspension of natural laws familiar on earth. For two billion years, life on this planet has been conditioned by those laws, and all evolution has been determined by them. In gravity, where everything has weight, skeletons are needed for rigidity and leverage, muscles are needed for any sustained motion, and a circulatory system is required to pump blood against gravity. Arms, legs, fins, and cilia have all been developed for locomotion in gravity. Now, in weightlessness, all the effects of gravity vanished, and with them many of the reasons men are the way they are.

Before Skylab, nobody had known much about the long-term effects of weightlessness on man, and it was the purpose of the project to find out what they are—together with assessing what sort of useful work man might do in space, and how he might comfortably live in his new environment. Before the end of the decade, NASA plans to launch a sort of combination rocket and airplane, the space shuttle, that should make getting to space and back a lot easier than it has been and about a fifth as expensive, assuming a certain traffic. Although the shuttle will have many uses, what NASA clearly has in mind is the later construction of a big permanent space station;

indeed, the shuttle's very name implies another terminal, aside from its space ports on earth, for it to fly to and from. A year after the Skylab missions were over, NASA sponsored a conference at its Ames Research Center outside San Francisco to consider the matter; the conference recommended that the United States look into building a huge space station that would house ten thousand men, women, and children, along with shops, schools, and industry, who would grow their own food and mine whatever metals they needed from the moon. Although no women, let alone children, have flown on American spacecraft, Skylab turned up nothing to preclude them in the future; there has already been one Soviet female cosmonaut, and NASA may include women on the shuttle. For long-duration missions with large numbers of people aboard, NASA already regards mixed crews as essential, provided there is an equal enough ratio to avoid the kind of barroom brawling that occurred in the old West, where women were in short supply. And, in fact, the conference at the Ames Research Center decided that the colonization of space should follow the pattern of the old West, which was first explored by small groups of men; and then by relatively young settlers, many of whom may have brought their wives with them; and finally, as more women followed, by a population whose makeup was the same as it was anywhere else. Indeed, NASA likes to present space stations as a sort of replacement for the old West, not only because space might provide a new source of raw materials and energy, but also, as the Ames conference saw it, because it provided "a way out from the sense of closure and of limits which is now oppressive to many people . . . in a world which has lost its frontiers."

In this regard, the first thing NASA had to find out was whether men were capable of spending long periods of time in space, and the indications were not good. Already, after previous American space flights, the longest of which—Gemini 7, which orbited the earth in 1965—had lasted only fourteen days, the flight surgeons had noticed certain adverse changes in the astronauts when they returned. They had lost muscle tissue and hence physical strength, and they had lost calcium, and hence rigidity, from their bones. Their cardiovascular systems were weaker. Their body fluids had been redistributed from their lower to their upper parts; and they had also lost a substantial quantity of liquids, including blood, which now contained diminished amounts of red cells, electrolytes, and hormones. The doctors had no idea whether these changes were the beginnings of trends that would continue at a steady rate of decline as long as an astronaut was in space; or, if the changes did level off at a point that was safe in space, whether that point would prove dangerous after the astronaut's return to the ground. One thing seemed certain: the longer a man was in space, the worse condition he was in when he came back. Three Soviet cosmonauts who had been in space for eighteen days in 1970 had been so debilitated when they landed that they had had to be carried from their spacecraft. Three other Soviet cosmonauts who, in 1971, had been in space for twenty-five days—the longest of any before Skylab—had died upon their return to earth under circumstances American flight surgeons had not felt were satisfactorily explained. Yet the first Skylab mission was to last twenty-eight days, the second fifty-nine, and the third, originally planned for fifty-six, would be extended to eighty-four. Beforehand, the commander of the second Skylab crew, Captain Alan L. Bean, had said, "We really don't know what's going to happen to the guys in the long term, and finding out is probably, in my mind, the single most important thing that we've got to do in Skylab." Though doctors and astronauts at the Space Center went around reminding each other that people were always apprehensive about new experiences —at the time of the first railroad trains, they jocularly told each other, some worriers had actually predicted that the speed would cause passengers to be unable to breathe—they were apprehensive nonetheless.

The symbol of one of the three Skylab missions was an adaptation of an anatomical drawing by Leonardo da Vinci of a man, his arms and legs outstretched, marked off with a square and a circle to show the proportions of his body and the length of his reach, a picture that suggests the analytical study of man; and perhaps at no time in history has anyone examined man so *exhaustively*, at least, as NASA did aboard Skylab. If a biologist wants to study an organism, he will subject it to a variety of conditions, but until very recently man has been studied in only one environment; by putting him in another, for as long a period and in as large a craft as Skylab, no one was quite sure what would be learned. The emblem of another of the Skylab missions included a test tube, which in a sense is what Skylab really was. Like all experiments, Skylab was very highly controlled. The men had to report every mouthful of food they ate, and keep track of every urination and bowel movement. Never before had so much data been assembled about a group of men for so much time. If Skylab was a test tube, the three crews were tiny colonies isolated in it, to be observed and manipulated by the computers and men at Mission Control at Houston. They were kept to a rigid timetable, and almost every minute was planned for them by the ground. What nobody foresaw—though it may have been one of the most important results of the experiment—was that some of the astronauts would rebel.

The third and last crewmen didn't always wake up refreshed in the morning, and sometimes they weren't eager to pop out of bed. Flight controllers and others at the Space Center, who had never been faced with reluctant astronauts before, openly talked of them as being lethargic and negative. They worried about them, and wondered whether there was anything in the strange alchemy of space that had changed their characters. They were unaccountably irritable, even when they were getting dressed in the morning. They complained that there weren't enough changes of clothes—and in this instance they were quite correct, for they were able to put on new ones only about once a week. But when they pulled new socks and underwear from their clothing bundles, and when a dozen other items floated out as well, filling the bunk rooms with a flurry of clean laundry, they bitched that the bundles had been packed too tightly. Then they grumbled that they didn't have a great enough variety of clothes; there were just trousers, which were golden brown, and turtle-neck T-shirts which were also golden brown—they had elastic cuffs so that legs and sleeves wouldn't ride up in weightlessness. (In spite of astronaut complaints, though, the clothing was quite versatile; if the men planned to spend much time in the docking adapter, which could be chilly, they pulled on golden-brown jackets, but if they planned to remain on the lower deck, which was warm, they unzipped the lower half of their trouser legs to convert them into golden-brown shorts.) The clothing was golden brown because that was the color of the fire-resistant material they were made of, and the crewmen didn't like that either. "I just get tired of this darn brown!" said Edward G. Gibson, the science pilot, who was thirty-eight years old and who, since becoming an astronaut in 1965, had gained a reputation as a natty dresser. "I feel like I've been drafted in the Army; this darn brown gets pretty obnoxious after a while. I'd like to get some different color T-shirts. . . ." Gibson, a slight astronaut with sharp features who was a civilian and a physicist, and who had a square jaw that apparently never stopped moving the whole time he was in space, was perhaps the contrariest, bitchingest astronaut that ever departed vertically from Cape Kennedy, and his two crewmates were in the same category. In addition to Gibson, the third crew comprised Lieutenant Colonel Gerald

M. Carr, its commander, a Marine with a wide, friendly face; and Lieutenant Colonel William R. Pogue, the pilot, a thick-set Air Force officer with a leathery but gentle face. Carr and Pogue grew thick, revolutionary-looking beards aboard Skylab, which the flight controllers in Houston could see on television growing thicker by the day, and which, combined with the blistering language from the space station, made them uneasy. The remarks of all three members of the third crew continued to have a barracks-room grumpiness from the beginning of the mission to the end. And like many barracks-room revolutionaries, the third crew was frequently quite accurate and penetrating.

As the astronauts got dressed, they patted their pockets to make sure they contained a number of items they would be needing—scissors, checklists, penlights, pencils, all sorts of odds and ends. As Skylab was so big that any object that wasn't held down could easily get lost, the astronauts had a great many pockets in their sleeves, chests, flanks, and running all the way down their legs; when these pockets were full, the crew looked like football players. "I'd. like to have a couple of garments around here which don't have these blooming pockets, just for comfortable, casual wear," Gibson grumbled. The men had to make sure that the flaps on the pockets were snapped shut. Sometimes the flaps popped open and the contents floated away. The third crew bitched that the pockets were too small for the items they were supposed to contain, and that therefore the flaps didn't close properly. "The pockets that are designed for the scissors don't have a flap long enough to cover the scissors," Carr, the commander, griped one morning after he had gotten dressed. "The one designed for the flashlight has the same problem. And so you end up having to put other things in those pockets, and you put the scissors and the flashlight in other pockets. In fact, the way it stands now, I have my flashlight in my scissors pocket; I have pencils

The Third Crew: Gerald M. Carr, commander; Edward G. Gibson, science pilot; William R. Pogue, pilot

in my flashlight pocket; and I have my scissors in the upper right pencil pocket. And every time I raise my foot to tie my shoelace, I jab myself in the groin with the scissors."

Carr, an easygoing Californian of forty-one, had some reason to be impatient. He had been an astronaut for seven years, and this was his first space flight—he was, in fact, the twelfth out of his particular class of astronauts, a group of eighteen who had been appointed in April 1966 to go into space. In contrast, both the other Skylab commanders, Captain Charles Conrad, Jr., of the first mission, and Captain Alan L. Bean of the second, had not only been in space several times, but they had both been to the moon. Astronauts regard the moon as the ultimate trip. Had the Apollo program continued for three more missions, as it had been originally scheduled to do, Carr would most likely have been aboard Apollo 20 instead of Skylab 3, a fact that sometimes made him a little regretful, particularly as the booster they were inside might have been the one to send him there. During his long wait to fly, he had worked hard on other astronauts' moon trips—he had helped with the lunar module that Conrad and Bean later landed on the moon, and he had been capsule communicator when their spacecraft, shortly after lift-off, had been struck by lightning. None of the third crew had ever been in space before; just prior to being launched, Carr told a reporter, "I'm sure that our lack of experience will be an embarrassment at one time or another in the mission," and he was right. Though Carr, Gibson, and Pogue would go on to be one of the most productive and interesting crews ever to go into space, where they also would remain longer than any other crew, their mission, which was to be the last in a spectacular series of space flights going back to the Mercury program, would have a certain *fin de siècle* quality.

The astronauts put on their shoes last of all. This was

the hardest part of getting dressed, for without gravity they had to force themselves to bend. "Up here, you've got to use those stomach muscles to pull your foot up close to you to tie your shoes or work on a sock or something like that, and it really gives them a workout," Carr said. These would be the only ones of the astronauts' muscles that would be *stronger* after their sojourn in space. As they bent, their feet rose from the floor, so that they sometimes did back somersaults before they got their shoes tied. Their shoes had canvas tops, like gym shoes, but they had aluminum soles coated with rubber to which they could fasten one of two different varieties of cleats. The floor in many parts of the spacecraft was made of a triangular metal grid, and without the cleats that the astronauts could attach to it, they would have been forever floating away. One was a triangular cleat that exactly fitted the triangles in the grid, and hence had to be attached carefully; when it was twisted into place, though, it held the astronaut absolutely firmly. The other was a smaller round cleat with a flange that the astronauts could slip in and out of the grid more easily, though it wouldn't hold them as securely. It looked like a mushroom, and that is what the astronauts called it. They were meant to try both varieties and decide which they liked best; in fact, they would spend more time fiddling with their cleats than with any other piece of equipment. One astronaut actually wore both cleats simultaneously for a while—a triangle on one foot and a mushroom on the other. The cleats were heavy; and as the astronauts left their bedrooms, they had to be careful not to let their legs flail around lest they club something or someone.

One morning during the first mission, while the first crew was still getting dressed, the wall phone next door in the experiments room buzzed, and the commander, Conrad, a bald man with a gap between his front teeth, floated out of his sleeping compartment to answer it. In contrast to the members of the third crew, who were nonheroes, Conrad was the stuff of which astronautical legends were made. He and his crewmates had saved Skylab. Through an error in design, just after the space station had been launched, the wind generated by the craft's speed had blown off its insulation, taking with it one of a pair of solar panels that supplied most of the space station's electricity, and pinning the other solar panel to the space station's side. With only the overhead array of solar panels—the ones that looked like a rotor —left to power the space station, and without its insulation, the temperature inside had soared; the heat had caused certain materials to vaporize; and to guard against these gases poisoning the astronauts, the flight controllers at the Mission Control Center at the Lyndon B. Johnson Space Center outside Houston had had to change the cabin atmosphere several times, venting it overboard and replacing it from tanks in the workshop as though they were rinsing a bottle. All in all, the first crew had had a frustrating time. When Conrad and his two crewmates, Commander Joseph P. Kerwin, the science pilot, and Commander Paul J. Weitz, the pilot—all Navy officers —had arrived after a delay of eleven days, Conrad had piloted the command module close to the pinned-down solar panel, while Weitz, standing in the open main hatch, had tried unsuccessfully to cut it loose with a sort of pruning hook. After that, they had attempted to dock with the space station, but had been unable to do so until the tenth try, after they had opened the command module's docking hatch, removed the docking probe, and repaired it. The next morning when they'd gone inside, they found the workshop's temperature was 125°, and in places it was 150°. To Conrad, who had been a Navy pilot, it felt like the engine room of an aircraft carrier. He, Kerwin, and Weitz had had to spend most of their time during the first few days in the docking adapter, where it was cooler, making only short forays into the workshop—just, Conrad had thought, like ships' engineers, who frequently came topside to cool off from overheated engine rooms. Though Conrad was one of the oldest astronauts—he had been born in Philadelphia forty-two years earlier and had graduated from Princeton in 1953—he fortunately proved to be in about the best physical shape of any of them. During his short stays in the workshop, Conrad, accompanied by Weitz, a muscular astronaut with a perpetual grin, had managed to push the parasol, which had been specially prepared in Houston, out an airlock hatch. Kerwin, a reserved, thoughtful doctor in the Navy medical corps, who normally smoked a pipe but could not do so in the space station's oxygen-rich atmosphere, had watched from a window in the command module as the sunshade had slowly emerged and then popped open like a big golden umbrella. After that, the temperature had begun to drop, though it had been a week before it fell into the seventies. The space station was still underpowered, though, and Conrad and his crew had had to turn out lights behind them wherever they went. Conrad had felt like a mole. At the end of the second week, Conrad and Kerwin had managed to go outside and free the solar panel that was pinned to the side of the workshop; without the extra power it provided, all the Skylab missions would have been severely restricted. Engineers and astronauts in Houston were so impressed with Conrad's handiness that after he returned they and other admirers presented him with enough hammers to fill a bushel basket. Even though they had saved Skylab, the first crew had been up only for a month, half of which was spent under emergency conditions; accordingly, it had never had a chance to settle in and sample life in space, as the two later crews would.

With the wall phone in the experiments room still buzzing, Conrad quickly twisted his triangles into the grid on the floor beneath it. (If he had been in more of a hurry, he might simply have stuck a toe in the grid, something the harried first crew did so often that the toes of their shoes wore through; as a result, the second crew's shoes had to be hastily outfitted with toe guards.)

"Houston, Skylab," he said into the speaker.

"Good morning, Skylab. We're in radio contact over Carnarvon for the next five minutes," the voice of the capsule communicator, or CapCom, also an astronaut, boomed through the space station. There were several squawk boxes spotted around the walls, so that the astronauts could talk to the ground, and to each other, wherever they happened to be. If too many of the squawk boxes were on, they picked up sound from one another, rebroadcasting sound and picking it up again so that it blended into a constant raucous squeal. The astronauts found it easier to use radio headsets concealed in each of their soft, black-and-white pilots' hats.

"Good morning," the commander answered.

"You sound awful cheerful this morning," the CapCom said. Conrad was indeed one of the most cheerful astronauts; when he had landed on the moon as commander of Apollo 12, he and his partner had filled the airwaves with jokes, howls of laughter, and snatches of song.

"Yeah, we got a good night's sleep last night," Conrad answered. "And I think we're just getting more used to sleeping up here. Everybody has tromped out. I've got to see where they are." He called to his two crewmates, Kerwin and Weitz. Conrad had to shout, because in the space station's low-pressure atmosphere—a third what it is on earth—sounds didn't travel readily. Fifteen feet was about the limit of a loud speaking voice, and consequently the astronauts were often quite hoarse.

Having located the others in their sleeping compartments, Conrad reported back, "Everybody is awake, and

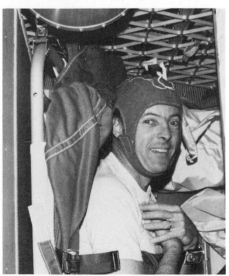

The experiments room on the lower deck, where the phone was buzzing; Conrad emerged from the door to the sleeping compartment (right) to answer it. The door on the left leads into the waste-management compartment or bathroom.

Joseph P. Kerwin, the science pilot of the first crew, still in bed wearing the bunny cap that monitored his sleep.

the science pilot is just angling himself from his bunny cap." The bunny cap, which no astronaut liked to wear, contained a number of electrodes which the science pilot wore during the night so that his sleep could be monitored. Kerwin, minus his bunny cap, came on the radio to give a brief report about how he had slept. "I woke up about three hours into the sleep period, checked the brain cap, and I had good contact on the electrodes," he began. Conrad interrupted, "What he really did was wake up three hours into the sleep period and go slip the bunny cap on the *pilot's* head." The CapCom said he bet Weitz, the pilot, had appreciated that. Weitz, who had come on the radio phone, told the CapCom that they were all going back to bed now, and the CapCom replied that this wasn't their day off. There would be no such merry exchanges between the ground and the dour third crew; when they talked of going back to bed, they meant it.

As the period of radio contact was about over, the CapCom said hurriedly that he would pick them up again about forty-five minutes later. Skylab, which orbited the earth once every ninety-three minutes, was in touch with Mission Control only about a quarter of the time, far less than any previous manned spacecraft. NASA's network of ground stations was mostly close to the equator, from which previous flights had never strayed far; but Skylab's orbit was so steeply inclined to the equator—fifty degrees—that the craft was only apt to be in touch with a ground station for a few minutes about twice each orbit, and frequently less than that. The sporadic scraps of conversation were annoying to both the astronauts and the men on the ground. Only during the same short periods of radio contact could the flight controllers get the telemetry, which told them the condition of the spacecraft, and it always took them a few minutes to analyze it; consequently the CapCom frequently had to cram everything he had to say into the last few minutes of a contact—occasionally he was even cut off before he was finished. Because the astronauts were in touch with the ground so sporadically, they talked to the CapCom mainly about immediate operational matters; whenever they had anything less urgent to relay, such as a long report on an experiment or comments about the space station, they dictated it into a special tape-recording system. Later, when Skylab was over a ground station, these tapes were transmitted to the ground (or, as space engineers say, "dumped") very rapidly over a radio frequency called the "B channel," to be distinguished from the "A channel" over which the astronauts talked with the CapCom. One astronaut said later that he actually *preferred* the B channel, for he felt he was talking to only one person —or maybe even just to himself. Indeed, the third crew, whose communications with the CapCom would be terse at times, made up for their taciturnity by being positively garrulous over the B channel. Gibson, who had started out with the intention of keeping a diary, gave it up and used the B channel instead. Some of the astronauts thought the B channel would be kept private, and consequently they were quite candid when they used it.

When he was through talking to the ground, and before he could move on, Conrad had to twist his shoes out of the grid. He had to stop and remember whether he was wearing the triangle or the mushroom, for they disengaged differently: the triangles twisted out and the mushrooms lifted out, and if he did the wrong thing, he could trip. Later this almost happened to Carr, the commander of the third crew, who dictated on the B channel: "I have gotten so used to using the triangle that I found that I would twist my foot when I wanted to disengage the mushroom, too. Of course, rotating the mushroom doesn't do anything, and you end up with your foot still locked in. And I almost twisted an ankle, stepping off in my usual manner as if I had a triangle on, and finding out that my foot was still anchored." Fortunately, in space, where nobody weighs anything, bones do not break easily.

One morning during the second mission, the second crew's commander, Bean, who had just finished his own before-breakfast chat with Mission Control, pushed himself with one hand off the wall by the squawk box in the experiments room and drifted slowly toward the bathroom. No astronauts had traveled this way prior to Skylab because their spacecraft had been too confining; in fact, they had hardly been able to move about at all. Bean and the other Skylab astronauts didn't think much about their novel form of transportation once they got used to it. "Seems to me that after eight to ten days, I decided that it just seemed natural to be in zero gravity. . . . It's not uncomfortable; it's pleasant," said the science pilot on the second crew, Owen K. Garriott, a slight man with a precise, neatly trimmed mustache. The experiments room was the biggest and most cluttered on the lower deck; it contained the ergometer, an exercising machine that looked like a bicycle; the lower-body negative-pressure experiment, an aluminum barrel big enough for an astronaut to get into; and the human vestibular function experiment, a rotating chair like a dentist's. It would have been easier if Bean had simply *floated* head first over all this bulky equipment; however, the astronauts always moved about on the lower deck, where the ceiling was low, in an upright manner, very much as they might in a room back on earth—as Bean had in fact done, by shoving himself sideways off the wall, so that he glided as though he was sliding across the room on ice skates. Sometimes they actually *"walked,"* which they managed by pushing against the ceiling, and then stepping with both their arms and legs.

There was a reason for their caution. Although there was no gravity in space, the workshop was designed as though there was—that is, there was an artificial sense of up and down, or what the astronauts called a "local vertical," provided architecturally by the fact that there was a definite floor and ceiling; and the astronauts felt most comfortable when they, and the room, were the same way up. Getting out of kilter was both exhilarating and worrisome—as Garriott found out one day when he took a walk on the ceiling of the experiments room. "It was a strange sensation. You see brand-new things," he said into the B channel. "You just have no idea how cluttered up the ceiling, which is now the floor, has become. Wires and cables and everything else tumbling all over. And it's really like a whole new room that you walk into. It's a fascinating new room. It's a pleasant psychological sensation just to see it with the lights underneath your feet, and it's just an amazing situation to find yourself in." Previous astronauts had never had this feeling because in their small craft they had always known exactly where they were. The three science pilots, Garriott, Gibson, and Kerwin, seemed to think more about the curious perspectives of weightlessness than did the others; and Kerwin, who spent the most time trying to analyze the problem, came up with an answer that interested the flight surgeons on the ground. "It turns out that you carry with you your own body-oriented world, independent of anything else, in which *up* is over your head, *down* is below your feet, *right* is this way and *left* is that way; and you take this world around with you wherever you go," Kerwin told the B channel, as though he really were Leonardo's man inside the circle. This phenomenon had never been apparent on earth, where a man's own local vertical was overridden by his sense of balance. Kerwin's discovery explained why an astronaut felt so odd if he got upside down or sideways in a room, because if he had two local verticals to worry about, his own personal one and the room's, he would naturally become confused if they got out of line. Gibson noticed that he could even get lost inside the space station by entering a room a different way up from usual. "I can move into a given room sideways or upside down and not recognize it; or perhaps I would recognize it, but I could not feel at home in it," he told the B channel during the third mission. In space, the astronauts had lost their sense of balance *completely*, a fact they repeatedly demonstrated in one of the medical experiments on the experiments room floor, a revolving chair that an astronaut strapped himself into and then rotated in at up to thirty times a minute, almost as fast as a phonograph record, while at the same time moving his head up and down, backward and forward, and from side to side. If a man tries this on earth, he quickly becomes perspiry, dizzy, and ill, because the motion disturbs the fluids in three semicircular canals in his inner ears, which determines his sense of balance. In space, no astronaut had trouble doing this, and if he was blindfolded, he didn't even know he was turning. The flight surgeons were perplexed, for there should have been some response to such a workout of an astronaut's sense of balance, which operates by the effect of centrifugal force on fluid in the inner ear rather than by the effect of gravity. Something was going on that the flight surgeons did not understand. Though Pogue, the pilot of the third crew, found the situation "good for kicks," it was clearly more trouble than it was worth, and as a rule most of the astronauts preferred the "floor."

Consequently, the astronauts felt most at home in the shallow space of the lower deck, which also contained the most domestic rooms, the bedrooms, the experiments room, the bathroom, and the wardroom; it was very much like an ordinary apartment on earth (except for the fact that it was circular and the rooms were more or less wedge-shaped) ; and most of the astronauts found it was easier to get around there than in other parts of the space station. But as Bean floated upright across the experiments room toward the bathroom, he could look through the triangular-grid ceiling overhead, which served as the

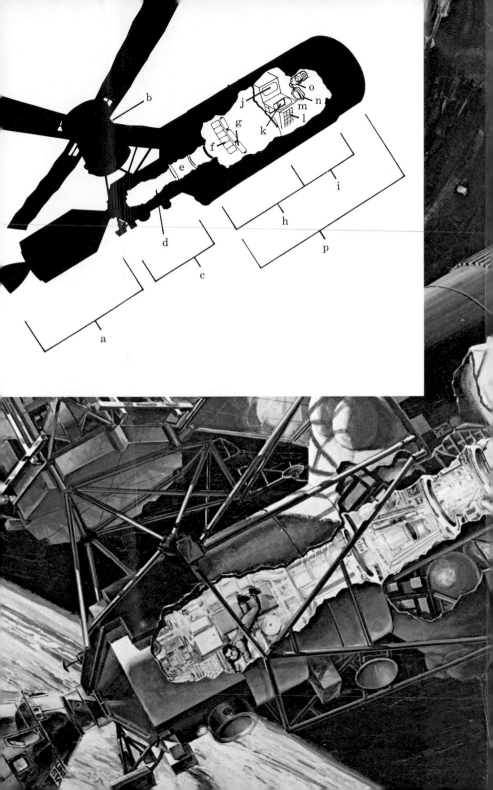

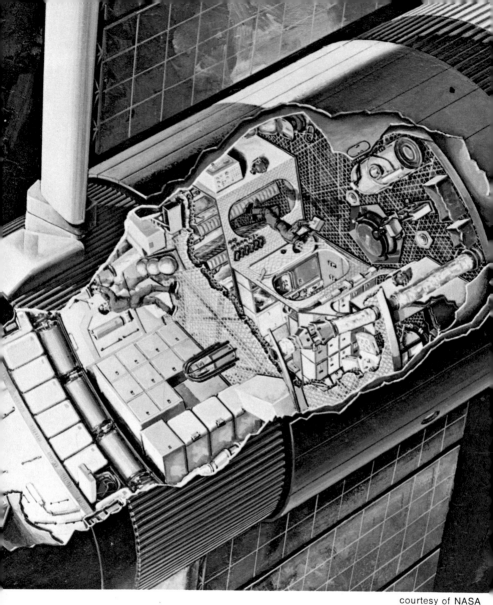

a	Command Module	j	Sleeping Compartments
b	Solar Telescope	k	Waste-Management Compartment (Bathroom)
c	Multiple Docking Adaptor		
d	Solar Console	l	Wardroon
e	Airlock Module	m	Experiments Room
f	Ring of Storage Lockers	n	Trash Airlock
g	Ring of Water Tanks	o	Lower-Body Negative-Pressure Experiment
h	Upper Deck		
i	Lower Deck	p	Workshop

floor of the deck above, and make out the rounded dome far away at the forward end of the voluminous upper deck, and the hatch at the dome's center that led on into the tunnels of the airlock module and the docking adapter. Up there, where the astronauts would have to get about in less terrestrial ways, existence would be more challenging, complex, and exhilarating.

Bean, a merry-looking man with receding hair, glided into the bathroom—a bit of privacy the Skylab astronauts enjoyed over previous ones—through a normal rectangular doorway, quite different from the round hatches he would have to dive through head first in other parts of Skylab. He pulled the bathroom door, a little like the folding one of a telephone booth, closed behind him. The bathroom—or the waste-management compartment, as NASA called it—was a small room about the size of a similar compartment on an airplane, and Bean bobbed about in it as though he were riding over a rough air pocket. This happened because the compartment lacked the triangular gridwork, which was hard to keep clean; instead, the floor was of sheet metal. "You just ricochet off the wall like a BB in a tin can!" one of the third crew would complain later. Bean, uncomplainingly, wedged himself into a corner and opened the door of the cabinet that contained his hygiene kit. Immediately his toothbrush, toothpaste, razor, extra blades, and everything else inside came floating out, and he had to corral them all back in again. This was a phenomenon, common to cabinets all over Skylab, that the engineers in Houston had dubbed the "jack-in-the-box effect." Bean later lined the bottom of his cabinet with Velcro—the plastic adhesive with interlocking bristles—then painstakingly glued small patches of Velcro to his toothbrush, his toothpaste, his razor, and everything else, and in this way he kept them laid out in an almost military fashion, a trick he recommended to all future space station occupants. Bean was an efficient, good-humored man—as Conrad's part-

Jack R. Lousma, the pilot of the second crew, ricochets off the bathroom wall, his right hand extending toward the astronaut's sink.

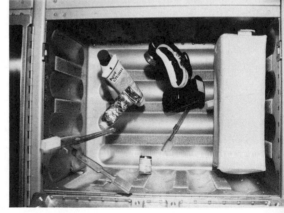

The "jack-in-the-box effect"

ner on Apollo 12, he had laughed as much as his commander on that happy trip to the moon—and so were the other two members of the second crew: Garriott, the science pilot, and Major Jack R. Lousma, the pilot

The second crew was the antithesis of the third. Several months later, when Gibson, the science pilot of the third crew, tried to use the Velcro Bean had, he would botch the job and accuse NASA of providing a cheap variety. He would blow up, "This Velcro doesn't stick! Let me resolve here, you should have put a couple of wads of bubble gum aboard, and they would have worked just as well. That Velcro is lousy! We ought to get that good Velcro and stop piddling around with this stuff! You gotta find another way of doing it; this way you lose whatever you happen to be holding down." Bean and the other members of the second crew had no trouble with the same Velcro; they never expressed themselves with the third crew's blend of ridicule, exasperation, and plain hostility.

The second crew was almost a continuation of the first. Both groups were happy with the space station and well disposed toward the ground. There were a number of close connections between the first two crews: the two commanders had not only gone to the moon together but they had known each other in the Navy, where they had been roommates; and it was Conrad, in fact, who later persuaded Bean to join the astronaut corps. Bean, in turn, had recruited Weitz, the pilot of the first crew. Weitz had known the pilot of the second crew, Major Lousma, a Marine, when they had been together at the U.S. Naval Postgraduate School in Monterey, California, during the early 1960s. (In the opinion of some at the Space Center, the fact that nobody in the third crew had been in the Navy, and hence had missed out on these connections, may have further isolated those astronauts from earlier ones.) Bean and his two crewmates were clearly the favorites of Mission Control. They were obliging and cheerful and did everything they were told; and when they had finished their work, which was always performed expeditiously and efficiently, they were apt to ask for more. When the third crew was flying, Mission Control would look back wistfully at the second mission and wonder what had gone wrong.

The second crew of astronauts seemed happy even as they went about washing up. They couldn't bathe in any conventional way because water wouldn't stay put in any sink or tub. They washed their hands by spraying them with water from a valve recessed into the wall; the water squirted all over, though. Suction in the drain was supposed to take the water away—the bathroom was forever making gurgling sounds like the lavatory of an airliner —but the drains didn't always work. The astronauts' suggestions for improving matters ranged all the way from providing a sort of autoclave they could stick their hands into entirely, and where they would be squirted in safety, to a sandblasting device that would do away with water altogether. To wash themselves more thoroughly, the astronauts rubbed themselves down with wet washcloths, of which hundreds were neatly rolled up in cabinets and color-coded for each astronaut with red, white, or blue bindings. The astronauts tethered their washcloths or towels by pushing the corners into slits in soft black rubber knobs that held them as securely as towel racks on earth; each knob had a color-coded picture of Snoopy next to it so that there would be no mix-up. A great many items in the spacecraft were similarly labeled, so the astronauts did not want to look at the comic strip "Peanuts"—which is the subject of cult-worship at the Space Center—for quite a while after they got back. One morning, Lousma, the pilot of the second crew, a fair, youthful-looking man from Ann Arbor, Michigan, who was perhaps the most exuberant of all the astronauts, televised Bean, his commander, as he gingerly squirted some water onto a washrag and rubbed himself down. Lousma was also the best at giving running commentaries over the B channel of events as they happened inside the space station, as though he were a sports announcer on a college radio station. ". . . OK, he gets the towels and washcloths right there to clean up with," he said. "OK, he's got the red one. There's a little hot-water dispenser right there. That's our sink. Here's a bar of soap; has a little metal piece in it to hold it to the magnet that's in there. He squirts a little water on his rag. See the little drops of water flying all over. Water doesn't go down here. Now he puts his washrag in the washrag squeezer to get the water out of it. Holds it out and hangs it up to dry. That's a little rubber-type device to stick things onto, just like he's doing there; [it will] hold down washcloths or towels or clothes to keep them from floating all over."

Every couple of weeks the bathroom doubled as a barbershop, providing Lousma with further material for his breezy broadcasts: "Yeah, well, what's going on here? We've needed haircuts for quite a while, so that's what's going on. Here's Captain Alan Bean who is the leader of this mob, and here is the distinguished professor Owen Garriott trimming his hair. Doing pretty well, too. He's a-flicking off the hair with that little blower up there. I might show you the tools of the trade used in Skylab, much like you might use on earth—a little comb with a razor in it. It seems to be working quite well. It's just a plain old hair comb. We also have some bandage shears, which come out later when we have to patch him up. Well, I can see it's not going to be a professional job, but there's no waiting and the price is right. . . ."

"Well, here we are again, space fans. We thought we'd drop back and see how it goes after a while. One of the advantages, of course, of having your hair cut in zero gravity is that you don't have to sweep the floor. The hair doesn't fall down. It doesn't even get on your shoulders. It gets up in the air sometimes and you just vacuum it

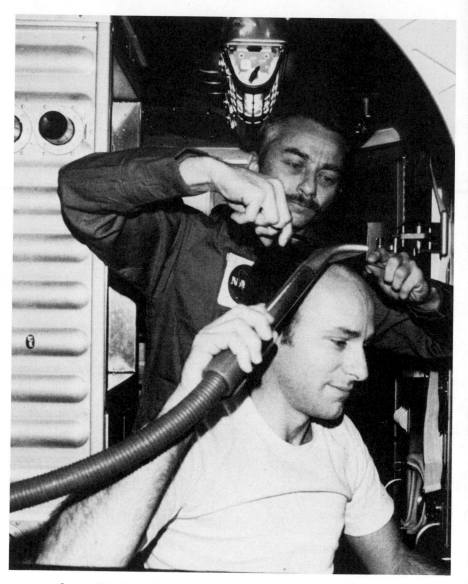

Owen K. Garriott, the science pilot of the second crew, gives a haircut to his commander, Alan L. Bean, while Bean vacuums the snippings.

up like this. That's what I'm doing—catching this loose hair with the vacuum cleaner. Doing a nice job, Owen. You might wonder why we chose Owen to do this job."

"Yeah, why?" Garriott asked.

"Well, we figured you could always trust a barber with a mustache. Well, we'll come back and check on them a little later and see how they're doing. You know there aren't many folks that get their hair cut at eighteen thousand miles an hour. . . ."

The *third* crew of astronauts had a miserable time in the bathroom, beginning with the moment they first entered and tried to see themselves in the mirror. As Carr tried to see his reflection, he had trouble making out more than a metallic smudge; the mirror, made of aluminum because no glass was allowed in the spacecraft, was dull. "I think there are better metal mirrors available than what we've got," he grumbled. As commander, Carr normally reserved his complaints for more momentous matters, but an additional aggravation now was that the only light in the bathroom, a single overhead bulb, was so situated that it was impossible not to cast his shadow on his reflection in the mirror. When he finally got himself in focus, his face was invisible in an angry swirl of reddish-brown beard and shaggy hair. Even the beardless astronauts, such as Gibson, sometimes had trouble recognizing themselves. Their faces were puffy, because without gravity to pull their blood and other body fluids to their feet, the fluids migrated to their upper bodies, including their heads, and stayed there—a phenomenon that affected all the crews, though the third made the most fuss about it. "Our faces are puffed and slightly chubby and red, resembling what you would see for someone hanging upside down in one G [or gravity], only not quite as pronounced," Gibson reported one morning. "We also have the feeling of a stuffy nose. Along with the red face, we have slightly bloodshot eyes. We expect it's for the same reason: high blood pressure in the upper extremities." The astronauts constantly felt as though they were coming down with colds. They sneezed a lot and sometimes had nosebleeds. They kept nasal sprays and tissues handy. Pogue, the pilot of the third mission, suggested that tissues be packed in all stowage areas aboard future space stations. Gibson, though, didn't like the tissues. "I would like to have a couple of plain old handkerchiefs around here," he grumbled once. "I'm not quite too sure why we have to go around plucking tissues all the time."

The redistribution of fluids toward the astronauts' upper bodies because gravity no longer pulled them down was a complex matter that the flight surgeons in Houston wanted to know more about. Six months after the end of the Skylab program, Kerwin, the first crew's science pilot, would say at the Skylab Life Sciences Symposium held at the Space Center from August 27 to 29, 1974, "It really is extremely clear to an individual when he is in weightlessness that rather profound changes are rapidly taking place in his body. One feels this strange fullness in the head and this sensation of having a cold and the nasal voice, and one sees the puffy look on the faces of his fellow crewmen. . . . One can almost see the fluid draining out of the legs; one looks at his partners, and their legs are getting little and skinny like crows' legs, and one knows that one's physiology is changing." Periodically, the members of the third crew measured their girths at their calves, thighs, waists, chests, biceps, and necks; they found that the circumferences of their legs became thinner by about an inch, while those of their upper bodies increased. This redistribution of fluids was responsible for more of the trends the flight surgeons had noticed, and had a greater impact physically on the astronauts, than any other single effect of weightlessness. Probably it accounted for the astronauts' loss of balance, which should have operated to an extent even in space, by increasing the amount of fluid in their inner ears.

Also, most of the blood pooled in their chests, where it reduced the breathing capacity of the lungs. Some of the fluid wound up in the astronauts' backs, where it was absorbed by the spongy disks between the vertebrae, making the disks thicker so that the vertebrae were pushed farther apart. As a result, the astronauts gained an inch or two in height—very much the way people on earth who have been in bed all night are a little taller in the morning. (On the ground people lose their extra inch rapidly, and this was to happen to the astronauts, too, after they returned home.) As they were also losing weight in space, at the same time as they were getting taller, the astronauts found that their trousers were noticeably short and loose.

The third crew did not have a merry time splashing itself with water the way the second crew had. Carr complained that the soap was like dog shampoo. Pogue, the pilot, bitched that the towels—which were made of a synthetic material that was highly fire-resistant—were "sort of like drying off with padded steel wool." Gibson griped that "the fire-prevention guys really got away with something when they made us go with that kind of material; I don't think it's absorbent enough, and I think it's too hard." Their absorbent qualities were so bad that whenever the astronauts took a shower—which they did very seldom because the shower stall took a long time to erect—they had to dry themselves with the vacuum cleaner, which they said didn't work very well, either. In particular, they hated the toilet, which was recessed halfway up the wall. "Well, I don't know how that was designed," said Pogue, "but I'm sure it wasn't by anyone who took a crap and noticed his posture." The third crew sometimes gave the impression that it was occupying a totally different space station from the earlier crews. Much later, after he was back on earth, Pogue, who was the earthiest of all the astronauts, was asked why he, Gibson, and Carr were so much more critical

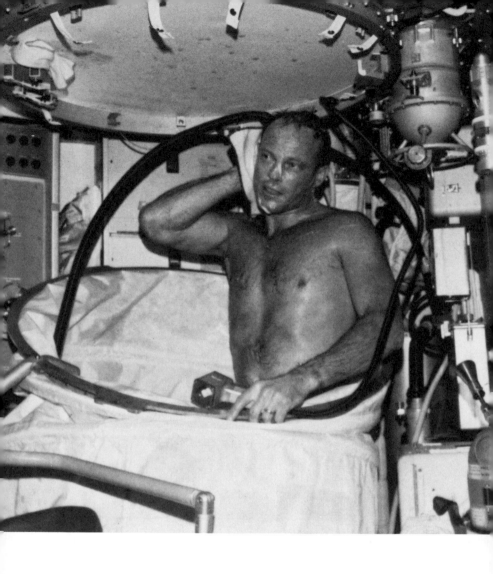

Lousma takes a shower. He doesn't like the soap any better than the third crew did.

than the others. "The purpose of these debriefings was to be hypercritical!" he protested. "If you try to be nice, it's a crummy debriefing. If you try to be polite or mince words, you emasculate the truth. If you're mealy-mouthed, you don't give a true picture. You gotta say it like it is." Astronauts rarely had before.

There was more behind the third crew's testiness than the wish to provide the ground with conscientious debriefings. These astronauts' troubles had begun almost as soon as they had entered Skylab, when an incident had occurred that seriously strained their relationship with Mission Control and made them feel defensive from the start. Astronauts were apt to feel nauseous when they first arrived in space. Indeed, the second crew, which had gone directly into the space station as soon as it arrived, had promptly thrown up; all three astronauts had been ill for the next several days, and they had had to take it easy for several days after that. The flight surgeons did not know what had caused the second crew's nausea, but they believed that there was an analogy between seasickness and what had come to be called spacesickness, for both evidently had to do with the fluid inside the three semicircular canals of the inner ear that provides the body with a sense of balance. At sea, any roughness makes the fluid slosh around in a dizzying fashion, sending confusing signals to the brain that somehow causes nausea; and in space, although there are no rollicking waves, the lack of gravity causes the fluid, if not to slosh around exactly, then at least to behave differently. And also, in the first few hours an astronaut is in space, the liquids in his body are in the process of redistribution so that the fluid balance in his inner ear is undergoing a change; while this is happening—and before the fluid reaches a point at which it totally shuts off his sense of balance—there might be a period, some flight surgeons later speculated, when an astronaut is particularly susceptible to nausea. Though the doctors weren't sure of the exact mechanism, they suspected that plunging directly into the large interior volume of the space station had made matters worse for the second crew, and they had resolved to keep the third crew inside the small command module for about twenty-four hours after its arrival so that the astronauts could acclimate to weightlessness before going inside. The astronauts of the first crew, who had done this, had had no trouble with nausea; the reason, though, might equally well have been that those men were so busy with repairs that they didn't have time to be sick.

In spite of the flight surgeons' resolve, Mission Control had allowed the third crew to enter Skylab thirteen hours early. It had been hard to hold the astronauts back. They were excited, and they had plenty to do. They had brought up with them in the command module about a ton of extra food, clothing, film, and experiments so that their mission could be extended from fifty-six to eighty-four days, and they had been anxious to get the material stowed. At first the work of settling in was easy because being weightless was novel and exhilarating, but after a while the astronauts had found that getting about became tricky and troublesome. It took a while to get the hang of moving around in space; for several days they overshot and crashed into things. Also, they became a little queasy. When Carr tried to get an antinausea pill out of a plastic bottle, it didn't just fall out when he turned the bottle upside down, and this upset him more. "The trouble with this little pill is you got to play games in trying to get just one of them out of there," he said irritably, giving it a vigorous shake. "Oh!" he said as it floated away. Gibson helped him chase it. When Carr caught it, Gibson suggested that he put one or two more in his pocket in case he needed them later. In the voluminous interior of the space station, the men were beginning to feel like passengers on an ocean liner who had done too much too soon. Carr, and particularly Pogue, were having more serious symptoms of what flight surgeons referred to as "stomach awareness." Looking at Pogue, Gibson remarked, "Ol' sweaty-palm time there." Pogue threw up. According to regulations, the astronauts should have informed Mission Control, and they should have freeze-dried the vomitus and brought it back to earth for medical analysis. They decided to do neither. Astronauts often play down any adverse effects of space travel because the more congenial an environment they can make space appear to be, the more likely NASA is to receive funding for future programs. "We won't mention the barf," said Carr, the commander, "and we'll just throw it down the trash airlock." Gibson concurred, saying, "They're not going to be able to keep track of that." Pogue said, "It's just between you, me, and the couch." And Carr said, "You know damn well every manager in NASA, under his breath, would want you to do that." As it turned out, the matter was between the astronauts, the couch, and a tape recorder for the B channel which subsequently relayed the conversation to the ground. Skylab was bugged as thoroughly as the Nixon White House.

The entire dialogue, neatly typed and mimeographed along with the rest of that day's B channel transcripts, had been on the desk of many NASA administrators the next day. The crew had been reprimanded; it was the first time astronauts had ever been reproved publicly during a flight. In the opinion of one of the capsule communicators, Dr. Story Musgrave, himself an astronaut, the incident and the reprimand doubtless affected much that went wrong later. Afterward, he felt, these astronauts' conversations with the ground became terse and frequently querulous. They began making errors, such as venting overboard an antiseptic for their drinking water, and they set some switches wrong when they made some initial tests of their equipment. They dropped further and further behind their schedule. However much the nausea episode had demoralized them, though, something else was wrong as well. It took the flight controllers, flight planners, and flight surgeons at Houston half the mission to figure out what the real trouble was and to set it right.

Washed and shaved, the third crew floated over to the wardroom for breakfast. It had Skylab's only big window, a round one 18 inches in diameter; it was covered now so that television and movie cameras on wall brackets could film the astronauts as they dined. Later, psychologists and engineers at the Space Center would study these and other films to see how quickly the astronauts adapted to weightlessness in such matters as their efficiency in moving about or handling things; and also to assess the habitability of the space station itself. The latter was sometimes a problem. Though the wardroom was the astronauts' favorite spot in Skylab, it was disconcerting in many ways. Here and there, stretched across the walls, were long elastics with hooks at either end called bungee cords. From time to time, without any warning, one came loose and whipped across the room as if it had been released from a slingshot; in weightlessness, it kept on wriggling and squirming for several minutes, the hooks thrashing dangerously. The third crewmen worried that they would lose an eye. The astronauts, who had nothing in the way of shelf space, tucked loose items behind the bungee cords; one, taut across a cabinet door, had in back of it a sheet from a checklist, a slip of paper with a memo on it, and a science fiction paperback by Ray Bradbury. Sometimes an astronaut grabbed onto a bungee cord in error, mistaking it for a handhold; and when that happened, it came off in his fist while whatever had been stuck behind it floated away. The wardroom walls were lined with cabinets; some of the cabinet doors were brown and some were sort of tan off-white; the browns had been chosen by a New York design firm in hopes of minimizing the pallor of the astronauts' faces; the firm had felt that the men's increasing whiteness, the longer they were incarcerated in the space station, might alarm them. Gibson, who experienced no such alarm, wished for a livelier color scheme, just as he had with the clothing— "something in a light blue or light green or something." Garriott, the normally reserved science pilot of the second mission, had objected even more strenuously. "It seems to me that the color arrangement that we've got in here might very well have been designed by a Navy supply department or something with about as little imagination as anybody I can imagine; all we've got in here are about two tones of brown, and that's it for the whole blinking spacecraft interior," Garriott said in answer to a long series of questions that each astronaut was required to answer over the B channel. Though they had been given a battery of light meters, sound meters, and aerosol-particle meters to help them describe such matters as color, lighting, noise, smells, and other items of interior decoration, most of them were quite capable of delivering their opinions without recourse to these devices. Generally speaking, the science pilots tended to be more discriminating about their surroundings than the military crew members; indeed, Bean, a Navy captain who may have been the easiest to please of all the astronauts, said he found the browns quite fetching.

As the dining table took up most of the space in the wardroom, Gibson and Pogue had to squeeze around Carr, whose place was nearest the door, as though they were in a crowded restaurant on earth—something that puzzled the psychologists and engineers at the Space Center who studied the film, because it would have been much easier for the men to soar over the table. They didn't, it turned out, because they feared they might become entangled in strings overhead that tethered checklist booklets such as one for operating their electric food trays; also, they no more wanted to get upside down in relation to the wardroom than they had in the experiments room. During the first mission, Kerwin, the science pilot, had found himself standing inadvertently on the wardroom ceiling. He said later, "I immediately became concerned that the table was on the ceiling, and the books of checklists that normally hung down from the ceiling seemed now to be growing out of the floor, like plants, the red and white pages fanning open like flowers. I wondered what they were doing there. It was bothersome—an Alice-in-Wonderland feeling." Though it was good for kicks, it was clearly a sensation to be avoided by astronauts about to dine. Accordingly, Carr, Gibson, and Pogue fastened their triangular cleats to the grid under the table, where it was a slightly different size from the grid in the rest of the space station. Carr, who seemed to have more trouble with his cleats than the others, managed to fasten himself in all right, though he had trouble unfastening himself later. "When you lock yourself into the grid in the wardroom, it's fine, but when you get ready to unlock yourself, it releases you before you get your triangles fully unlocked," he said on one such occasion. "And then you find yourself skipping along on one foot, ricocheting off the walls or the overhead, trying to reach down and get your triangle popped back into the neutral position so that you can stick it into the grating somewhere else." If they had wanted to, they could have sat on chairs of a sort, which pulled out from under the table. They were little more than parallel bars, though, which they could have tucked their legs through as if they were sitting on rail fences; in weightlessness, astronauts hardly needed anything larger. Sitting, though, required bending at the waist and staying bent; and as doing this put a constant strain on the stomach muscles, which were already quite sore from all the bending they had done that morning when they were getting dressed, the crew preferred to stand. Anchored now around the table, they crouched slightly, for in weightlessness, when

an astronaut relaxed his muscles, he assumed a somewhat fetal stance that engineers at Houston called the
"neutral G position." An astronaut's arms tended to bend
at the elbows. Consequently, all the astronauts thought
the table was too low. The best table for space might
turn out to be one that is considerably higher and also
has a slanted top, like a drafting table.

For the first time, astronauts in space were eating
more or less as they might at home—not only were they
seated, or crouched, around a table, but they had silver
as well. (Previous astronauts had eaten out of tubes or
bags.) For the third crew, though, eating breakfast was
a fight all the way. The wardroom table was little more
than a pedestal that supported three food trays, and the
trouble began the moment the men unfastened the lids,
which even the second crew complained were held down
by what Lousma, who at times sounds like a member of
the third crew, called "the most miserable latch that's
ever been designed in the history of mankind or before."
Lousma, though, wasn't in the same league as the third
crew when it came to criticizing the table. "Well, I
wouldn't want the people that designed that table to do
anything else," Pogue fumed later. "I think that it's such
a lousy job of design that I wouldn't want to have those
same people work [on the table further] because all
they'd do is just make a bigger and better white ele-
phant." What bothered Pogue in particular, once the lids
were off, was that there was no secure place to put his
knife, fork, or spoon. "We need better ways of restraining our silverware," he said. "I've put rubber bands
around my tray and I hold my stuff down that way. . . .
But while looking out the wardroom window, we've
kicked the utensils off the table, and I've got a spoon
stuck on the [ventilating screen] upstairs right now."

Meals in Skylab often tended to be rather Rabelaisian,
from the point of view of squalor if not abundance. The
night before, each astronaut had fitted his breakfast food,

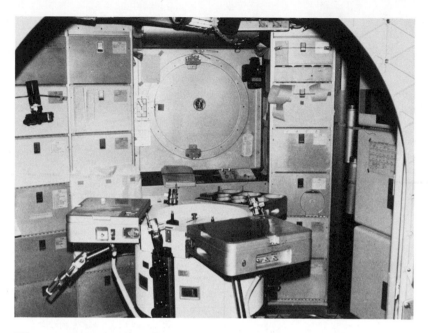

The wardroom table, little more than a pedestal that supports three food trays, stands in front of the round window. Above left, a camera is stuck behind a bungee cord stretched across a cabinet door.

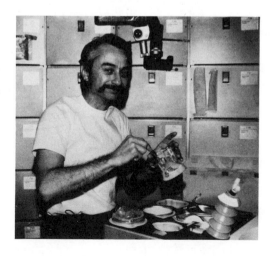

Garriott eats breakfast out of a plastic bag. Plastic squeeze bottle, for juice, is at the right.

which came in cans, into round holes inside his food tray. In the morning, removing the tray lids was like opening Pandora's box, because most of the food cans were too small for the holes they were in, and they floated out. The astronauts had to catch them and wedge them back in. Carr suggested that future space stations should have dispensers for food, as though the wardroom was an Automat. Each astronaut had selected what he wanted to eat before he left the ground, but he was very apt to regret his choices—and his choices were repeated every six days. "Beef hash for breakfast!" Lousma had grumbled one morning. "I've asked myself every six days, whenever it turns up on the menu, 'How come I picked beef hash for breakfast?'"

The astronauts had to be careful as they opened the cans, for the pull tabs were sharp. No delicious smells of sausage, bacon, eggs, or hash rose to greet them. For one thing, the food inside the cans was still contained in plastic bags called wet packs. For another, the food was tepid, for the food trays, which were meant to heat the food in time for meals, didn't work well. In any event, smells, just like sounds, didn't travel readily in the space station's low-pressure atmosphere. And even if none of the above had been the case, the congestion in the astronauts' heads prevented them from smelling very much and, what was worse, from tasting very much, either. This was probably the trouble with the hash, which had tasted all right to Lousma when he had sampled it in Houston. All of the food tasted bland. To make their eggs taste better—or taste at all—the men had to put lots of salt on them. They squirted the salt out of little dispensers, for it was in solution, since the dietitians at the Space Center had thought loose salt crystals would be hard to handle in weightlessness. "Message for the food folks," Gibson dictated to the B channel. "The salt dispensers are not working out too well. The nozzles for the valves at the top of them seem to plug up with dried salt, and in order to make them work, we have to squeeze the bag. Unfortunately, we squeezed one bag too hard, and we have been picking salt off the walls of the wardroom for two days. I think we need a better method." Gibson at times seemed positively to revel in his difficulties. Dissolving the salt was still probably the best method, as some pepper the second crew brought along in particle form floated about inside its box so that it was almost impossible to shake any out; and what did come out, instead of settling on the eggs, swarmed off in a cloud, diffusing throughout the workshop. The second crew, who had given more thought to the food than the third crew, and who had managed to eat more neatly, had brought up with them a variety of condiments and spices, including horseradish and Tabasco sauce, but nothing seemed to give the food much zip.

Astronaut meals may have had a rather bland and processed quality to begin with because most of the food, and all the drinks, were dehydrated and had to be reconstituted with water—in some cases this was done the night before, and in others it was done at mealtime. The water came from what looked like a dentist's fountain at the center of the wardroom table. The astronauts consumed a great deal of water in their rehydrated food and drinks; and indeed most of them made a point of taking a sip whenever they were passing by the wardroom. This was because, with the redistribution of fluids in their bodies, the extra blood that now pooled around their hearts created the false sensation that their bodies contained too much liquid, and this in turn triggered some mechanism that depressed their sense of thirst; if they didn't force themselves to drink, they could become quite dehydrated. Each astronaut had his own Snoopy-coded water hose that was metered so that he could keep track of his water consumption; he had to be careful not to chip a tooth on the hose's metal nozzle. The button that operated the fountain was stiff, and sometimes when an astronaut pushed down too hard and wasn't anchored to the floor, he himself went up in the air. The crews recommended that any water fountains on future spacecraft operate with a pistol-grip trigger that would keep an astronaut on the floor by containing within the handle the force of any button-pushing. Water sipped with such difficulty, though, might not have done them much good, for another effect of the extra blood now pooled around an astronaut's heart was to give him a false sense of satiety, and to cause him to eliminate what the body took to be excess fluid. Most of the astronauts on earlier flights had weighed four to six pounds less after their return, and the flight surgeons attributed half of this to fluids lost in this fashion. It was along with these fluids that astronauts on earlier, shorter flights had lost electrolytes, hormones, and red blood cells; and the flight surgeons worried that if the astronauts continued to lose liquids on the longer Skylab flights, they would be in serious trouble. Electrolytes, for example, are body chemicals whose continued loss could lead to arrhythmias of the heart, such as the Apollo 15 astronauts had suffered on the moon.

At breakfast, the third crew used their hoses to add cold water to their powdered orange drink, which came in plastic squeeze bottles that expanded like accordions, and added hot water to their instant coffee. The water was filled with air bubbles because the air that had pressurized the water tanks was never able to float to the surface in weightlessness and consequently remained mixed in; the bubbles gave them gas. "I think farting about five hundred times a day is not a good way to go," Pogue declared. The bubbles in the water caused trouble, too, when the astronauts rehydrated their solid food, which was contained, inside the food cans, within the clear plastic bags called wet packs. When the astronauts injected these with the bubbly water, the wet packs were apt to explode and blast the food all over the wardroom, spattering the walls, the windows, the grid ceiling, and even getting beneath the grid floor. If the wet packs didn't explode, the astronauts cut them open with a pair of scissors tethered to a string overhead, alongside the checklists; another string anchored their vitamin pills. "As soon as you manipulate the bag the slightest bit, squirt! Out comes a squirt of food that goes lurching off into the workshop somewhere, and you have to go catch it," Carr said. Sometimes they had to spin the bag so that the food would go to the bottom; then they complained that their silverware was too short to reach all the way down without getting their fingers dirty. The silverware came in for a good deal of criticism. In particular, the men groused that the bowls of the spoons were too small, a feature that had been deliberately provided because in weightlessness liquids tend to cling together in globs, and engineers on the ground had worried that a conventional bowl would give an astronaut too big a mouthful. The astronauts thought otherwise. "We don't eat with tiny spoons on earth; why do they expect us to do it here?" a member of the second crew had griped; and a member of the third crew, never to be outdone by the second when it came to complaints, said, "The spoon bowls as far as I'm concerned are lousy. I would give them a rating of lousy on your scale, which is somewhat below poor." (The ground had given the astronauts a scale, ranging from "Excellent" to "Poor," for evaluating everything in the space station.) Still, they ate with the spoons in preference to the forks, because the food— sticky stuff, such as cold cereal gummy with sugar or meat and vegetables in cloying gravies—adhered better to the spoon bowls than to the prongs of the forks, which had less area, and which, as a consequence, received an even lower grade than the spoons' "lousy."

Keeping their food from flying off the silverware was a problem. If an astronaut at breakfast spooned up a bit of egg and then stopped his hand halfway to his mouth

—to ask someone to pass the salt, say—the egg would leave the spoon, and, continuing on its own trajectory, impact on the astronaut's face. "I find myself eating Japanese style quite often, getting my mouth down very close, and this just minimizes the possibility of getting food loose, because you get your mouth open right down next to the food, and you can sort of shovel it in," Pogue said during the third mission. Later, though, the astronauts learned to eat more gracefully. The trick was not to stop the spoon on its trip, so that the food wouldn't fly off. They developed a smooth, arclike motion, tipping the spoon slowly as it went so that it would always be directly in back of whatever was on it. They had to have their mouths in exactly the right spot and keep them open, for there was no way to stop the spoonful of food once it had started on its way. As they ate and went about their other chores, the cameras on the wall brackets whirred. The film, when it was shown back on earth later, was jerky, because very few frames were used per second; and as the astronauts chased after gobbets of food, or got egg on their faces, they looked a little like Charlie Chaplin—someone else who seemed uncomfortable when he was projected into an inimical world. Space had a way of turning these rather serious men into comedians.

The second crew had gone about their morning chores with considerably more efficiency and good humor than the third crew. After breakfast, the Skylab astronauts normally spent about fifteen minutes cleaning up and doing a variety of odd jobs before starting in on the day's experiments. Sometimes they had to make an inventory of their remaining food supply. "Now here comes locker five-fifty-four, so stay alert," Bean had said into the B channel one morning toward the middle of the second mission, as Garriott stood by to help. "Let's go with the big cans first: corn, corn, corn, corn, strawberries, strawberries. Try the next big can: applesauce, salmon salad, corn, white bread, white bread, corn, white bread, white bread, white bread. A lot of bread. Mashed potatoes, mashed potatoes, mashed potatoes. Six mashed potatoes. The guy that likes mashed potatoes has got it made. . . . Now here comes a light can. Nothing in it. We'll ask somebody to get rid of that. On the way down!" He tossed it to Garriott. "That's it, gang, for compartment five-fifty-four. Now let's go to five-fifty; let's go with the big ones first, gentlemen. Green beans, three; chicken and rice, three. Veal, veal, veal, asparagus, five; strawberries, five. Six corns and six turkey rice soups. Sounds like a loser. Strawberries, seven. A guy could live on strawberries up here. Strawberry shortcake, turkey and gravy, there's five turkey and gravies. One, two, three, four macaroni and cheese; one, two, three, four shrimp cocktails, four. And two pears."

After breakfast, Garriott, a tall, wiry astronaut who, though he was an electrical engineer, doubled as the second crew's physician, had opened up the medical dispensary. It was equipped for all sorts of emergencies, ranging from extraction of teeth to appendectomies to cardiac arrests, and flight surgeons on the ground were always available to give advice over television. Despite the curious physiological changes occurring to them in space, the science pilots never got much medical business; about the only times the television was used for this purpose was on one occasion when a member of the first crew had an ear infection, and on another occasion when the members of the third crew had skin rashes. The second crew had no such problems, and indeed when Garriott had tested out the TV camera's diagnostic abilities by giving Lousma a medical examination with it, he was hard put to find anything to focus on. "At the moment we're looking with our close-up lens at Jack's [Lousma's] eyeball," Garriott said, camera in hand. "And I'm going to zoom in a little bit and give us a little closer look at some of the things that we can see. The little flecks in the iris can be seen, and the pupil. [Perhaps] the little bloodshot streaks are related to conjunctivitis. . . . Let's see if we can see back in the throat now. OK, come up a little bit, that's right, OK . . . OK, now we are shining our little flashlight back into Jack's mouth, and had we any throat problems, perhaps laryngitis or any sort of sore throat, we would be able to get the advice of a trained physician on the ground. . . . We've obviously got a pretty good view of the teeth, in case we should have a chipped one. See a filling there in Jack's left molar. Had that filling come out, we have on board all the necessary equipment to refill that tooth temporarily." Garriott, a literal-minded man, had a sort of deadpan humor; he may have been getting back at Lousma for his emceeing of the haircut. Not only was it hard to find anything the matter with the astronauts, photogenic or otherwise, but they *felt* fine, too, at least after an initial period of adjustment, which took about two weeks in the case of the first crew, four weeks in the case of the second, and a lengthy six weeks in the case of the third. Kerwin, the science pilot of the first crew, who as the only doctor ever to go into space had been most aware of the curious changes that were going on inside himself and his mates, had asked himself a long list of questions every morning after he woke up, such as, "How's my sleeping? Eating? Drinking? Eliminating? Any signs of stress anywhere? Am I making out all right, or am I paying some sort of a penalty?" The changes notwithstanding, Kerwin said later that he and his crewmates, Conrad and Weitz, felt downright "bouncy."

Bounciness notwithstanding, the flight surgeons in Houston still wanted the science pilots to continue their medical examinations. Garriott drew some blood with a syringe from each astronaut (something Lousma enjoyed even less than the medical examination) and whirled it in a centrifuge inside a cabinet on the lower deck. Later, the flight surgeons would examine these daily samples to follow the loss of red cells and to establish when that trend started and stopped, if it did. They would check on something else that they understood even less, for while the astronauts were in space, some of their red blood cells were changing shape; normally they looked like tiny red checkermen, concave on both sides; but as time went on a large fraction of them distorted, becoming thin, attenuated, and flat.

The astronauts never got very far from the medical aspects of life in space, even when they were cleaning up the wardroom after breakfast. It was Lousma's turn to tidy up, after Garriott had finished with him. For the Houston flight surgeons, he had to "weigh" the leftover food in the cans—something of a problem in weightlessness—but he accomplished it by putting each one inside a wire mesh box and then pressing a button so that the box rocked back and forth; the amount it rocked was an indication of the can's mass, a figure that the ground could use to determine its weight had it been on earth. (The same principle was used when the astronauts, also at the request of the flight surgeons, "weighed" themselves, which they did every morning by rocking in a contraption on the upper deck that looked like a rowing machine.) The amount of leftovers would tell the flight surgeons exactly how much the astronauts had eaten and drunk, and this information—together with their own weights and also daily samples of their excretions, which they would bring back to earth with them—would be used by the flight surgeons to study some of the other trends they were worried about, such as when the losses of hormones and electrolytes began, how much was lost, and when the losses stopped, if they did. While Lousma was packing away the astronauts' own wastes, he remarked that six days' worth occupied almost exactly the same volume as six days' food—a fact that made him toy with the idea of moving any food he didn't like, such

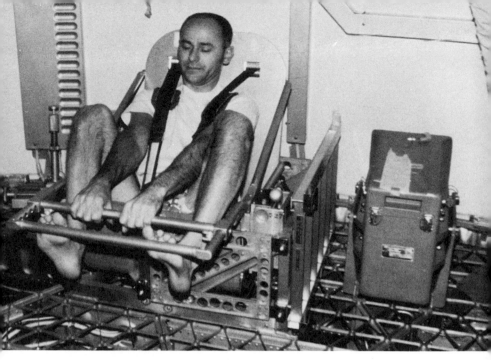

Bean "weighs" himself by rocking in a special chair that determines his mass; flight surgeons in Houston will later compute his weight.

Charles Conrad, Jr., the commander of the first crew, obligingly turns upside down so that Kerwin can examine his teeth at morning sick call.

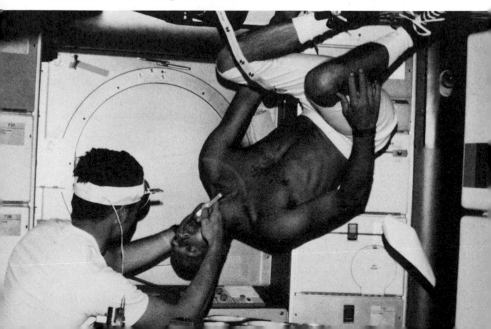

as the beef hash, directly from one container to the other, bypassing himself.

When Lousma was through weighing the food cans, he crushed them in a press to reduce their size and put them into a white trash bag with a silver lining—there were several of these bags, which the astronauts used as waste-baskets, fastened about the space station. Even though they had specially prepared antiseptic wipes for cleaning up, the astronauts preferred using their old clothes, and they were forever raiding their trash bags for cast-off shirts that could be ripped up. (As there was no washing machine aboard Skylab—a lack most of the astronauts felt should be made up aboard future space stations—they simply threw their laundry away.) Lousma remarked, in a motherly fashion, that he could always use a rag around a space station. After he had wiped up all he could, he went over the entire lower deck with a small vacuum cleaner. He had trouble vacuuming the grid floor, though. Most of the astronauts complained that the machine lacked suction—a curious complaint, in view of the fact that the space station was surrounded by a near-perfect vacuum; however, the vacuum cleaner was unable to tap it for suction, as other equipment did, lest too vigorous a housecleaning evacuate too much of Sky-lab's atmosphere. Fortunately, the space station never got very dusty because fans sucked the air through a couple of filter screens on the workshop ceiling. The circulation kept odors down, too; the space station, which Lousma once said sometimes smelled like a gymnasium, he admitted was normally quite fresh. The air-conditioning machinery made a whine that he said he found a relief from the quiet of space. Avoiding the various meters that had been put aboard to amplify his own senses, Lousma added, "The spacecraft just hums along with a very comfortable noise level; when the fans are not running, you miss them." The air was recirculated through a big beige pipe down the workshop wall and beneath the floor of the lower deck, whence it began its upward journey all over again. This constant upward current was an unexpected boon in housekeeping, for all sorts of debris wound up on the filter screens, so that all the astronauts had to do to keep the spacecraft spic and span was to vacuum the screens every day or so.

As Lousma went about tidying up, he stuffed into his lower right-hand pocket any items he came across that might be useful again, such as bungee cords that had come loose or bits of tape, but other odds and ends of junk that had to be thrown out he stuffed into his lower left-hand pocket, where they would remain until he changed his clothes. Then he would put his old pants and jacket, with their pockets bulging with flotsam, into a trash bag, and when the bag was full, it would be flushed down the trash airlock, which was in the middle of the lower deck in the experiments compartment. The trash airlock emptied into the garbage tank aft of the workshop; without it, the astronauts' rubbish would have followed Skylab—as indeed had been the case with the space capsule in Jules Verne's book *From the Earth to the Moon*, which was surrounded by its own nebula of empty wine bottles, chicken bones, and even the body of a dog who had died on the trip. The trash airlock, which stood in the center of the floor like a small garbage can with a long, leverlike handle, was complicated to operate, for the tank beyond was not pressurized, and consequently it had to be flushed without losing any of the space station's atmosphere. One night, an astronaut who was making his way in the dark back to his bedroom, unwittingly jarred the handle with his foot; it wasn't until the next morning that a gentle hissing sound revealed that the air in Skylab was slowly leaking away. Later, Bean, the commander of the second mission, attached a springy bungee cord to the handle to hold it down. Bean had a healthy respect for the airlock. He was terrified that it would break down; since there was no backup for it, if

Lousma vacuums dust from the filter screen on the workshop ceiling, all that is required to keep the space station spic and span.

it did break down the astronauts would soon be swamped in their own garbage. "If we lost that trash airlock, it would be one of the worst things that could happen," Bean said once. That had almost happened to the first crew when a big cannister got jammed in it; those astronauts had tugged away at the handle for an entire orbit before the obstruction had passed through. (In Houston, where technicians help solve flight problems by duplicating them and then trying to solve them, a group of engineers had jammed a similar cannister in a similar trash airlock in a replica of Skylab to see if *they* could find a way to dislodge it. They tugged away for an entire orbit, too, but without success.) Bean, who always seemed to have the greatest sense of responsibility for the space station, never let anyone but himself operate the airlock, and he advised Carr, the commander of the next crew, to do the same. "There's a lot of feel to this thing," Bean briefed Carr over the B channel. Bean opened the airlock's lid, put in the trash Lousma had collected, closed the lid, pressed a button to vent the air inside, and then —first wrapping his legs around the airlock lest he levitate into space—he pulled the handle hard. He could hear the trash carom off the opposite wall of the tank below with a hollow thud.

Fed and shipshape, the astronauts were usually ready to start work at about eight o'clock in the morning. There were four main varieties of experiments aboard Skylab: solar astronomy, earth resources, medical, and a miscellaneous group known as the corollaries, which included botany, astronomy, biology, metals processing, and some simple experiments called science demonstrations. As the corollaries could be done in less time than the others, the ground controllers who were planning the astronauts' schedule used them as a sort of padding to fit around the larger blocks of time the astronauts devoted to the solar astronomy, earth resources, or medical experiments. For

Carr, the commander of the third crew, takes Bean's advice and flushes the trash airlock himself.

example, all the astronauts spent large amounts of time every day pedaling the bicycle ergometer, a medical experiment, which they were not supposed to do within an hour of eating. "You got three guys up here who do not like to exercise after a meal," Carr complained after breakfast one day, when Mission Control, by mistake, had him pumping away on a full stomach. "There are some people in this world I guess who think it's neat to do that, but we don't. And we would prefer not to do so." Accordingly, it was useful if the astronauts had a few corollaries to perform while their digestive juices were flowing.

After breakfast one morning during the second mission, Garriott was scheduled to do some science demonstrations. Many materials took on new physical characteristics in weightlessness, and the science demonstrations, or "demos," as the astronauts called them, were intended to experiment with the different ways certain items behaved in space. The most basic experiments, of course, are frequently the simplest, as was the case with Galileo's experiments with gravity, which required nothing more than a few iron balls and a tower to drop them from. Men had had so little experience with lack of gravity that it was at Galileo's level they had to start.

Garriott, whose trim, precise mustache made him look prim, had all the dry enthusiasm of an old-time naturalist. He was forever urging his two crewmates to do more experiments, as though he were a science teacher trying to inspire a group of apathetic students. Garriott's commander, Bean, got along very well with him, even though he himself was a professional pilot, a group that at NASA had traditionally opposed scientists as astronauts. "There's a real advantage to having a scientist-astronaut on the crew!" Bean said after he, Garriott, and Lousma were back on earth. "Lousma and I tended to think generally about the same things. Our background's the same, our interests seem to be the same; but Garriott always had a different point of view on a lot of things. We would find out Garriott was doing something, or thinking something, and either he would tell it to the ground, or he might mention it at dinner; and we'd say, 'You know, maybe *we* ought to look at that; we haven't been doing it.' " Bean was a little in awe of Garriott, and perhaps of scientists in general. Differences between the scientific and nonscientific members of an expedition, as well as being valuable, could lead to trouble, and had done so on expeditions to remote areas on earth, such as those to Antarctica, where, cooped up for months on end in close quarters, scientists on one hand, and military men or engineers on the other, had sometimes formed cliques that were mutually hostile. This did not happen aboard Skylab, in part because all the astronauts, both scientists and military men, had trained together, and spent enough time doing each others' jobs, so that there was no real division between them (something that might not be the case on large space stations with more specialized crews). In any event, with a science pilot like Garriott aboard, there was no opportunity for the other two to lose interest. He was, after all, doing experiments that physics teachers have dreamed about for centuries but have never been able to do themselves. As Garriott went about the science demos, he seemed a little like a child—a rather dry, professorial child—absorbed in play. And this was proper, for weightlessness was still very new to Garriott; as he floated items in the air to see how they would behave, he was defining his environment as surely as a child who pushes his glass of water off the edge of his table.

First he put a plastic squeeze-bottle of water, shaped a little like a rocket, in the air and left it there. He spun it, so that it rotated around its long axis, in order to keep it steady. Rockets have also been made to spin to make them more stable, but in the early days of space flight many of them had tumbled out of control because the fuel inside them sloshed around; and indeed as Garriott's half-filled water container spun on and on as though by magic, it, too, began to wobble, and the wobble turned into a tumble so that it was revolving end over end like a propeller. Next, Garriott spun a flat disk in the air, and it continued to spin ethereally without a trace of a wobble. Every once in a while it drifted out of the range of the television camera; whenever this happened, he simply blew it back to the center of the screen, like a prestidigitator with a levitating ball. The disk, he said, was much more stable than the rocket-shaped water bottle—explaining, perhaps, why extraterrestrials come in flying saucers instead of rocket ships. He could make it wobble, though, by tapping it as it spun. He took out the wobble by touching it, as though he were smoothing a clay bowl on a potter's wheel. It would remain absolutely steady in relation to the stars regardless of what the space station did, just like the tiny spinning electronic gyros that were at the heart of the space station's navigational system. Like a juggler, Garriott picked his disk out of the air and replaced it with a bar magnet, a small cylinder of iron about the size of a pencil stub. He flicked it slowly into the air, so that it tumbled gently in front of the camera; instead of continuing to tumble indefinitely, though, the magnet slowed down and down until it was simply oscillating back and forth; then the oscillations became weaker and weaker until the magnet stopped altogether, almost with a snap, as it lined up firmly with the earth's magnetic field. If Garriott had allowed *it* to remain where it was for an entire orbit, it would have turned gradually, making occasional jerks as it followed the earth's magnetic fluctuations. Next Garriott produced a loose spring, like the child's toy called a Slinky, with a sturdy metal plate at each end. He stretched it and pushed it together again, like an accordion; when he let go, it wriggled on and on like a snake, and would have continued to do so indefinitely except that he snatched it out of the air, bent it double so that the plates on either end were touching, and then released it, causing an unceasing nodding motion as though it were a two-headed serpent in constant converse with itself.

The enthusiasm of Garriott and the other two members of the second crew for doing the science demos and all the other experiments aboard the space station did not fail to impress Neil Hutchinson, the flight director in charge of all three Skylab missions, a lean man who talked energetically and claimed he couldn't stand any form of disorganization. He would be the nemesis of the third crew. When these astronauts first arrived in the space station, they found in a teletype machine to which the ground transmitted messages a note that read, "JERRY, ED, AND BILL, WELCOME ABOARD THE SPACE STATION SKYLAB. HOPE YOU'LL ENJOY YOUR STAY. WE'RE LOOKING FORWARD TO SEVERAL MONTHS OF INTERESTING AND PRODUCTIVE WORK." Hutchinson, a gung-ho, no-nonsense engineer, clearly intended to keep the astronauts moving as fast and efficiently as possible from the start. His day was full of considerations like no-bicycling-until-meals-digested, which he called "constraints"; and there were hundreds of them. Scheduling was so complicated—doubtless foreshadowing problems to come on the even bigger space stations NASA contemplated for the future—that Hutchinson and the other flight controllers had to resort to a computer. "We're building a flight plan every day as intricate as a lunar outing, and we have none of it planned before we leave," he said after the second crew had returned but before the third crew had gone up. "We send up about six feet of instructions to the astronauts' teleprinter in the docking adapter every day—at least forty-two separate sets of instructions—telling them where to point the solar telescope, which scientific instruments to use, and which corollaries to do. We lay out the whole day for them, and the astronauts normally follow it to a T! What we've done is we've learned how to maximize what you can get out of a man in one day." Hutchinson and the other flight controllers admittedly took great pride in their ability to plan an astronaut's day to the last detail, a type of thinking that would soon bring them into collision with Jerry, Ed, and Bill. And this was bound to happen, for now that astronauts were no longer flying short bursts of missions but were in space to stay awhile, they expected to have more of a say in how they lived.

Garriott's and the rest of the second crew's zeal for practically everything aboard Skylab proved the third crew's undoing. Garriott had urged Bean and Lousma, his crewmates, on to greater scientific endeavors to such an extent that all three of those astronauts had spent almost all of their free time doing experiments. The eagerness and success of the second crew had convinced Mission Control that it had better quickly find additional experiments for the third crew to do. About the worst thing the flight controllers, who were known even around NASA as a "highly motivated group," could imagine was having three astronauts up in space with nothing to do. Accordingly, the flight planners asked NASA's scientists and engineers to come up with some fast proposals. The response was overwhelming; there were always many more ideas for experiments around NASA than could possibly go into space, and the third Skylab mission offered the last opportunity for such experiments for a long, long time. And Mission Control, which normally tried to keep the number of experiments on a mission to a manageable minimum, was itself asking for more. "I think the flight planners had a lot of stuff jammed down their throats," Pogue, the pilot of the third crew, bitched with some accuracy as he wallowed in the new experiments, which were by no means limited just to the science demos. Most of them had been added so late that there hadn't been time for the third crew to become familiar with them before launch, or even know what their purposes were. Pogue said, "Somebody thinks something up in an office, it sounds good, and then all of a sudden you find yourself trying to do it for the first time [up here]; never having done it before, you're gonna take probably four or five times as much time to do the task than the man who has been needling the flight planners to have it included said it would. 'Have 'em do my experiment; it only takes five minutes,' he says, and you end up taking an hour to do it." What Pogue had in mind was a new medical experiment that he had never practiced, and that required him to float a few yards above one of his crewmates who was stretched out on the floor, wearing only his shorts, and then to photograph him with infrared film that would show his circulatory system—the purpose was to study the redistribution of body fluids by the size of the blood vessels. Pogue, though, found that it was impossible to hold still enough, while he was floating in the air, to take pictures properly— something he believed would have been anticipated if the experiment hadn't been such a last-minute rush job. To hold himself steadier, he once wrapped his legs around a water tank, kicked off the valve, and almost flooded the workshop.

Normally, experiments aboard a spacecraft took years to plan and develop, and hundreds of thousands of dollars to build; the new science demos for the third crew—most of them simple and using very little equipment, but still double the original number—were thrown together at the Space Center in only a couple of months. "Most of them were not well put together on the ground, and pretty much depended upon the guy up here to do the job," groused Gibson, the science pilot, who on the third mission had to do most of the jury-rigging. Gibson was the antithesis of the rather courtly Garriott, who had always been polite to Mission Control, had almost always found everything in the space station to his satisfaction, and had always been ahead in his work. About all the two science pilots had in common was that they were the only civilians among the nine Skylab astronauts.

When they buckled down, though, the members of the third crew put in performances that were often more inspired than those of their more workaday predecessors. Indeed, as Gibson, and also Pogue, did the science demos, they appeared less to be experimenting than simply having fun; at such moments, they forgot their troubles with Mission Control. They experimented a good deal with water, which in space forms the most fundamental of shapes, a sphere. Next to the simple fact that all objects float in space, the fact that liquid takes on the form of a perfect sphere was the surest indication that the astronauts were no longer on earth. The reason this happens is that in the total absence of gravity other, lesser forces become dominant, which means that the surface tension of the water, not a strong influence on earth, in space governs its shape, evenly holding the water inside a perfect globe. On earth, no sphere of any material is perfect, because gravity always pulls it out of shape.

The water the astronauts used for their science demos was in a big syringe; when Gibson pressed the plunger, injecting water into space, a small bubble grew from its point. When the bubble was completed, Gibson jerked the syringe away; the water immediately took the form of a tiny planet. Sometimes he or Pogue floated several planets in the air together; some of them Gibson dyed red with grape juice or blue with iodine to make them more visible, so that they looked like Mars or Venus. Together, as they hovered slowly, they resembled a miniature solar system. As he herded them about, Gibson thought their simple circularity was beautiful. The balls of water always seemed to catch Gibson and Pogue unaware. Sharing what seemed to be a greater capacity for surprise than the more matter-of-fact Garriott had pos-

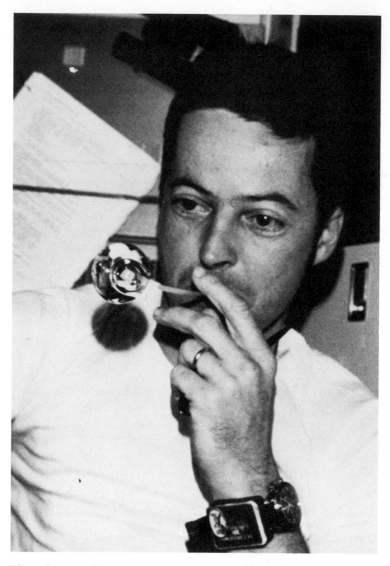

Kerwin, the first crew's science pilot, watches a ball of water, which he has just blown from a straw, as it floats away like a tiny planet.

sessed, Gibson and Pogue would do curious things with the little globes, which were slightly clammy to the touch and tended to cling, spreading across the hand like giant amoebas. Pogue once tried to herd two or more of them together to see if they would merge to make one big ball. With his black beard, he looked like a sorcerer's apprentice as he crouched intently over the levitating balls. "OK, here come two bubbles together," Pogue said, performing over the space station's television. "I think we're gonna make it this time. . . . Actually, I think they're touching and bouncing apart, which is sort of interesting. . . . There we go!" The two balls swallowed each other up, forming a single huge ball, just as though they were two balls of mercury that had come together. Sometimes Pogue jiggled the bubbles to make them oscillate; the motion had a symmetry unknown on earth, where spheres are imperfect. Pogue poked at a floating bubble with a long wire—he said he was tickling it—but each poke sent the bubble into such paroxysms of giggles that it rollicked off the screen. He set the bubble on a flat glass, where it was drawn down and out by cohesion with the glass to make a perfect hemisphere; when he tapped the bottom of the glass with a wrench, the bubble rippled like jelly, but with perfect rhythm. When he was through with the bubble, he attached it to a taut string along with some other bubbles; together they looked like beads on a necklace. Once, when one got away, it stuck to a wall like a blister; in the space station's dry air, it soon evaporated.

There was no end to the experiments the astronauts conducted with the globes of water. They stuck a syringe into a sphere and injected air so that it inflated exactly like a balloon, for the surface tension of the water was enough to hold it together and form a skin almost as stretchable as rubber. One day, as Gibson was inflating a sphere on television to see how the air would affect its oscillations, the bubble burst, spattering his face and the camera lens. "Well, you saw it, bubbling fans," he said, wiping himself with a towel. "This is one we'll have to try again." When he did, he had another problem, for now the air he injected, instead of making a single big pocket inside the sphere, made hundreds of tiny ones, so that the water looked carbonated; this time, though, when the water had all the air it could hold, instead of bursting, for every new air bubble he added, one that was already there was pushed out the other side of the sphere. The astronauts teased and tortured water into other extreme forms; like putty, water in space could be stretched long and thin. Gibson put together a small lathe out of spare parts in the spacecraft—this was the task that had annoyed him the most when he had complained of jury-rigging equipment—and then fixed a bubble of water between the lathe's two revolving points. While the lathe turned, he pulled the points apart so that the sphere between stretched into a long, thin cylinder. (The maximum length of the cylinder, it was later computed, equaled the circumference of the same amount of water had it remained a sphere.) The faster he rotated the lathe, the more the cylinder swung in a U like a long rope of putty; at length it broke, the water reforming into two bubbles, one at each point of the lathe. Gibson brought them together to form a single bubble once more; again he drew the bubble into a thin cylinder, but this time he rotated the two ends in opposite directions so that the cylinder was twisted, the way one might twist a rope of putty. As he increased the speed of the lathe, the water seemed to flow about, like the spiraling strands of some liquid rope, until at length it could keep up no longer and broke apart once more. In addition to stretching the water long and thin and twisting it, the astronauts could pull it wide and flat. Gibson placed a bubble on a wire loop and then gradually opened the loop to see how thin a sheet he could stretch it into. (He had had to make the wire loop himself, another cause for complaint.) While he was stretching the bubble out flat, but before he had pulled it into the thinnest possible sheet, there were any number of points where it still bulged, partially bubblelike, making a perfect lens. Perfect lenses would have great value on earth, where gravity makes them sag during production.

NASA had hoped the third crew would freeze one of these perfect lenses and bring it back to earth for study, with an eye to mass production of glass lenses in a future space station, but these astronauts never got around to doing this. NASA, of course, was forever on the lookout for profitable enterprises that might justify a big space station; two ideas that could not have occurred to the Skylab planners (though they were thought of at the conference at the Ames Research Center a year after Skylab ended) were the use of space stations to construct additional space stations and to construct huge orbiting satellites that would generate solar power to be beamed to earth in the form of low-density microwaves. Aboard Skylab, though, NASA was interested in the science demos and some of the other corollaries, to see whether there were simpler means of turning weightlessness to good account. For example, if molten metal formed spheres in space as readily as water, and was then allowed to cool into perfect ball bearings, NASA would be in the big money because machines using perfect bearings would run much longer before wearing out. (In theory they would be entirely frictionless and, therefore, last forever.) The first crewmen tried making these bearings in a small furnace up in the docking adapter, but they didn't succeed because there was no way of releasing the molten metal without creating a sort of stem, which, of course, ruined it as a bearing. NASA intends to try this experiment again someday.

Another possibly profitable attribute of space that the astronauts experimented with was the lack of convection. In weightlessness, hotter, lighter gases or liquids never rose to be replaced by cooler, heavier ones to make convection currents, which are the cause of most circulation on earth; the lack of them in space was the reason Skylab needed a powerful ventilating system. Indeed, fires the astronauts lit in a small furnace snuffed themselves out in their own smoke. NASA hoped to turn this phenomenon to financial advantage, for without convection to stir things up, the astronauts were able to mix certain alloys more perfectly in space than on earth. Also, totally new alloys could be made of materials that, because of their different densities, would have separated in gravity. The biggest potential money-maker of all, though, were crystals, which the astronauts grew more perfectly in space because gravity didn't deform them during their development. Not only did the crystals grow larger than they did on earth, but sometimes they took on totally new forms and properties; with them, the electronics for a computer that filled several cabinets could be reduced to the size of a transistor radio. And because these crystals were so small, NASA could transport them easily, and profitably, aboard the space shuttle.

When Pogue and Gibson were through with the science demos, they and Carr flitted through a big round hole in the center of the ceiling of the lower deck, where their relatively cramped living quarters were, on up to the upper deck of the workshop, a spacious room 21 feet in diameter and 20 feet high. The designers of Skylab had deliberately provided the astronauts with a variety of interior spaces to see which they liked best. The astronauts had caught glimpses of the upper deck all morning, any time they glanced overhead through the triangular grid, which was the floor of the deck above. In the upper deck, the astronauts were aware that they were inside what had once been a fuel tank in a way they weren't anywhere else aboard the cluster, for the cylindrical metallic walls rose straight overhead, bending together to make a

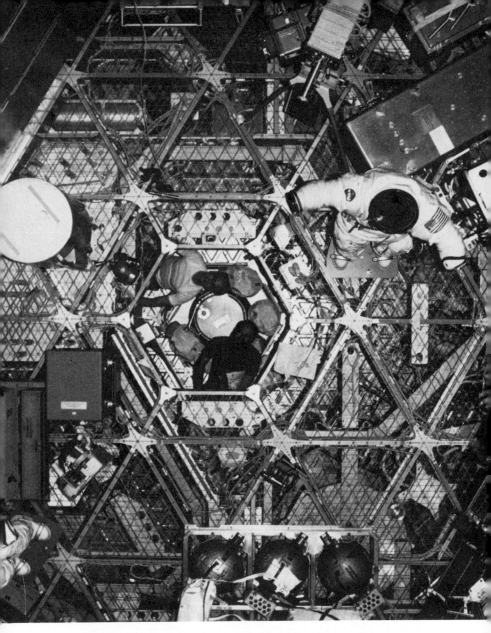

A view from the ceiling of the workshop's spacious upper deck, whose grid floor doubles as the ceiling of the more cramped lower deck; Gibson and Pogue, the science pilot and pilot of the third crew, are about to jump through the hatch from the lower to the upper deck. An empty space suit stands upright on the floor; air makes it move in a lifelike fashion.

rounded metallic dome at the top, leaving little doubt that they were inside a vast container for a powerful explosive. The only way out of the tank was straight overhead, through the small round hatch at the center of the dome. Shortly Gibson would rise vertically up the length of the workshop, in a sort of astronautical apotheosis, and disappear through it on his way to the docking adapter, where he would spend the rest of the morning operating the solar telescopes. Set into the wall just above the grid floor of the upper deck were a couple of small square airlock hatches that were used for pushing various scientific instruments outside the spacecraft; the one on the side normally toward the sun was of course occupied by the parasol the first crew had erected; its box, which was still attached to the airlock, stuck halfway across the floor. About 10 feet higher up the wall were two rings of big brown lockers, the main supply cabinets for Skylab, where most of the food was stored; above them was a ring of big blue water tanks. Overhead, dark gray against the white of the dome, were the two big screens that filtered the workshop air, and down the wall ran the thick beige ventilating duct that recirculated air to the lower deck, creating the constant upward current. In the workshop, so many ducts, wires, and hoses snaked up and down the wall that Bean, the commander of the second mission, who didn't know what they were all for and who was continually thinking up helpful tips to pass on to the third crew on the ground, advised his successors to find out before they came up.

The astronauts didn't feel as much at home in the upper deck as they had on the lower. Almost all of them preferred the small rooms of the deck below, where they felt more enclosed and therefore safer, because there was less risk of losing the sense of local vertical. Indeed, some doctors had worried ahead of time whether astronauts, without a sense of balance or any known replacement for it, would be able to orient themselves at all in a volume as big as the upper deck's. For all anyone knew, the astronauts might have reacted the way several minnows did, which Garriott had brought up with him in a plastic bag full of water that he attached to the workshop wall; bringing them had been Garriott's idea, for in addition to being the sort of scientific quartermaster of Skylab, he was also its best zoologist. At first, the fish evidently lost their vertical sense completely, for they just swam around and around in queasy-looking circles, first one way up and then another. After a while, though, they settled down and began swimming with their bellies toward the wall to which their plastic water bag was attached, as though they were aligning themselves visually with the bottom of a river. Presumably they were learning to use their eyes to replace their sense of balance. How quickly this could happen was demonstrated in another experiment with a special variety of spider that built a new web every day—the webs, of course, provided a graphic record of how fast the spider was learning to use its eyes to orient itself. The first web the spider built was an unrecognizable tangle, with the strands not even in the same plane; but by the third day, she was building webs that were in one plane and quite recognizable. Inside her circumferential web with its radiating strands, she looked as much at home as Leonardo's man—or as the astronauts themselves were becoming inside the nadirs and zeniths of their own body-oriented worlds.

For the astronauts, too, had learned to use their eyes to replace their inner ears as the link between their own verticals and that of the space station; and this, of course, was why they didn't flounder about in the upper deck, despite its volume. There, an astronaut could almost select, with his eyes, which vertical he wanted to follow, the room's or his own private one. "All one has to do is to rotate one's body to [a new] orientation and whammo! What one thinks is up *is* up," said Kerwin, the first crew's science pilot, who had discovered the phenomenon. "It's a feeling as though one could take this whole room and, by pushing a button, just rotate it around so that the ceiling up here would be the floor. It's a marvelous feeling of power over space—over the space around one. Closing one's eyes, of course, makes everything go away. And now one's body is like a planet all to itself, and one really doesn't know where the outside world is." Closing one's eyes, of course, could be dangerous, as. Kerwin inadvertently had demonstrated in a nonexperiment late one night, after everyone had gone to bed and the lights were out, when he was awakened by a radio call from the ground. "It was pitch black," he said later. "When I scrambled out of bed, I had no way of determining up from down; I had no visual reference in the dark. I had to turn on the lights, but I just didn't know what direction to put my hand in. So I had to *feel* things to orient myself—I had to use touch instead of sight—and everything felt different because I didn't know my relationship to them. It took me a whole minute just to get the lights on." The confusion passed as soon as he had lined himself up visually with the room's local vertical; indeed, when an astronaut's own vertical was lined up with that of his surroundings, the two seemed to click into place, like a compass needle onto magnetic north. "It's as though your mind won't recognize the situation you're in until it sees it pretty close to the right orientation, and then all of a sudden, zap! You get these transformations made in your mind that tell you exactly where you are," Gibson dictated into the B channel during the third mission. After receiving this message, a number of engineers at the Space Center, whose support of the astronauts was boundless, went over to the Skylab trainer—the replica of the space station—and had themselves hoisted about it upside down; they discovered that everything clicked back into place when they had returned to between thirty and fifty degrees of their normal vertical. Although it was easier for the astronauts to get out of kilter in the vastness of the upper deck than in the smaller rooms below, the local vertical was still pronounced enough so that they could always right themselves visually. They would not have this handy visual compass card in all parts of the space station, however.

Though most of the astronauts were stick-in-the-muds here, too, preferring to cling close to the floor, they did on occasion cut loose in the vast empty space. They were able to move about here with a freedom no man on earth, and no astronaut in space, had ever known before. All previous craft, both Soviet and American, had been so cramped that astronauts could always reach out and touch one, and usually more than one, wall; and on space walks, their space suits were as cumbersome and confining as straitjackets. The Skylab astronauts treated the upper deck as a sort of pool and gymnasium combined. Sometimes they threw a rubber ball, trying to see how many times they could bounce it around the ring of lockers high up the workshop wall—Weitz, the pilot of the first crew and the most athletic of the astronauts, bounced it 111 times before the ball died. Sometimes they threw Velcro-tipped darts, though at first the darts always missed the Velcro dart board; the astronauts, it turned out, were aiming high, the way they would on earth to counteract gravity, only in space the darts, of course, weren't dropping down; when the men aimed directly at the target, their scores improved. Sometimes they flew paper airplanes around the second deck; like the darts, they flew straight wherever they were headed, with the deadly accuracy of jet fighters.

On occasion, though, the crew preferred being projectiles themselves. In the upper deck, they often had to abandon the cautious, upright means of getting around they favored in the more earthlike rooms below. Instead, they shot about head first, like the darts and the paper planes, somersaulting at the last minute to land on their feet. They were able to soar as no human beings had ever done before, and they put on a sort of perpetual aerial ballet. Sometimes they had contests to see how many somersaults they could do. Conrad and the other members of the first crew had got so that they could jump from the floor to the ceiling of the second deck in a perfect three-and-a-half gainer, landing neatly on their feet. Men have always wanted to be weightless, and many sports are devoted to the illusion of total freedom from gravity: the racing driver on a fast turn, the skier on a jump, the parachutist during free fall, even a canoeist on white water. None of these sports could create the sensation of weightlessness for more than a few seconds, however. The astronauts were in this enviable condition all the time. "We were tickled to death!" Conrad had said after the first mission. "We never went anywhere straight; we always did a somersault or a flip on the way, just for the hell of it." Conrad was nimblest of all the astronauts, for he was shorter and more wiry than the others—an advantage, as in space long legs were a nuisance; they were forever getting bruised on hatchways. All the astronauts became more agile, though, for in one of the medical experiments designed to test their reflexes, when their heels were tapped with a rubber hammer, their feet kicked more than an earthling's; in weightlessness, their nervous systems sped up, like those of birds whose reactions must be swift.

Later crews became even more proficient at acrobatics; the tops of their shoes wore out because they kept brushing against the workshop walls the way swimmers would, with the tops of their feet. Sometimes they stood on each others' shoulders, breaking apart and remaining in midair in a way that circus performers would envy; or they soared around, three deep, on each others' backs, in a sort of human pyramid. The men of the third crew learned to slow their movements through the air while increasing the number of somersaults, so that Gibson—

lying on the floor of the second deck, hands extended over his head—could with a flick of his wrist send himself tumbling against a wall, give himself an additional shove with his feet, and do ten and a half somersaults before he hit the ceiling. The men had to be careful not to bang their heads, for metal equipment stuck out all over the walls. (The jutting equipment was sometimes useful to an astronaut for catching onto or steadying himself; one member of the second crew had remarked that wherever something stuck out, someone was always sure to grab onto it.) Each astronaut had a sort of crash helmet, called a bump hat, but none of them ever felt a need to wear one. The only mishap occurred during the first mission when Conrad, despite his nimbleness, had dislocated a finger by catching it in the triangular grid floor; Kerwin, who was a doctor, snapped it back into place. Most of the time the astronauts were very cautious, imposing on themselves a speed limit of two feet per second. "We were as careful as men walking on ice," Weitz, the pilot of the first crew, said. Like skaters, or divers, they were less interested in speed than in precision and a sort of swooping effortlessness. Lousma had liked to leap from the workshop floor, do a variety of gainers before reaching the ceiling, and then—just before he reached the center of the dome—straighten out as cleanly as a diver hitting the water, in order to disappear through the hatch without touching its side. They invented contests. The most challenging was to try to go all the way from one end of the space station to the other—from the trash airlock in the center of the lower deck to the command module, which, although some 90 feet apart, were in line of sight along the main axis of the space station. The idea was to make the trip in one smooth flight, without touching anything along the way. When the astronauts were in a hurry they could make the trip in fifteen seconds, but when they were trying to be accurate, they slowed down so that it took about forty seconds. Pogue,

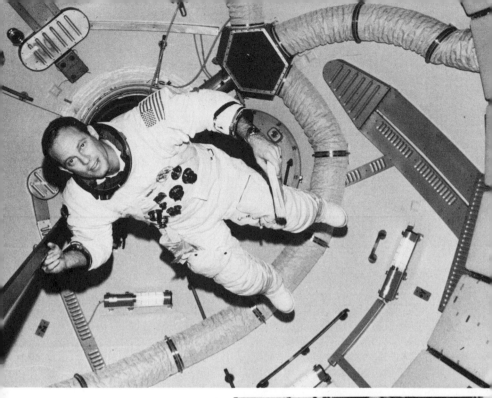

Acrobatics under the dome:

Lousma as Flash Gordon,

Bean does gainers,

Garriott marches and
does about-faces,

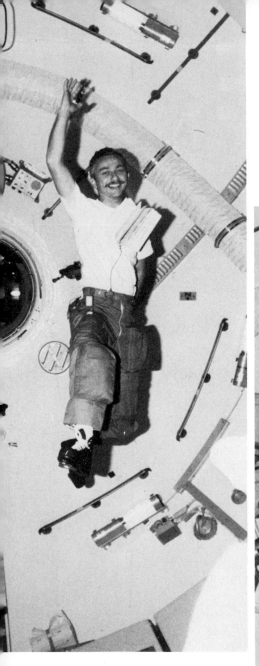

Lousma soars.

who sometimes felt he was the clumsiest of all the Skylab astronauts, was successful in only one out of twenty tries; there were four hatches on the way, and he usually banged his knees on all of them. Most of the others did better. Conrad had once made the entire trip holding the television camera in front of himself as he floated ethereally up the center of the upper deck, through the hole in the dome overhead, through a short tunnel into the airlock module, on through the docking adapter, and finally into the command module. The effect, as he threaded one hatch after another, was of a long, smooth dive through the bulkheads and corridors of a sunken ship.

Most of the astronauts recommended that any future space station should have ample interior volume for acrobatics. "You will need it as a place where people can get away from any claustrophobia which they might get in small compartments," Gibson said during the third mission. "At least, I feel that if I were penned up in [a small spacecraft] for months, it would begin to feel pretty much like a cell." The only astronaut who disagreed with Gibson was Garriott, the most practical, who had dismissed the upper deck as "waste space"; even so, he enjoyed tumbling about in it as much as anybody else. All the other astronauts agreed with Gibson that the upper deck had something of the value of an exercise wheel in a squirrel cage. Men who are in confinement for long periods—whether in prisons, submarines, or bases in Antarctica—develop various symptoms of isolation, chief among them a mutual hostility. In Antarctica, where the bases were much bigger than Skylab, the men's sense of release when one winter ended was so great that they walked out onto the ice, each in a slightly different direction, until they almost disappeared. Skylab's upper deck may have offered an even greater sense of freedom and release than the wide-open Arctic spaces, and it was built in. Perhaps, as a result, the Skylab astronauts got along with each other very well, especially the third crew, who may also have taken out its animosities on the ground. The second crew, which had gotten along so well with the ground, had had more trouble with each other, Bean's admiration for his science pilot notwithstanding.

The unconfining freedom of weightlessness was, in fact, not all that free, because the astronauts were now—as they never had been on earth—completely subject to the Newtonian laws of motion, as though they were themselves little asteroids in orbits of their own. Once they were in motion, the astronauts couldn't speed up or slow down, or change direction, without some sort of assistance. Pogue once tried to see if he could fly about at will, like a bird, by attaching big panels to his hands and feet and flapping them like wings—as though he were a flying machine out of Leonardo da Vinci's notebooks—but the wings notwithstanding, he floated inexorably onward until he crashed into a wall. This situation took some getting used to. Gibson found that if he allowed himself inadvertently to drift away from a wall, he had to float patiently along until he touched the opposite wall, a trip that once took him twenty minutes. (In such a fix, an astronaut normally asked a crewmate to give him a shove.) Kerwin had speculated about what would happen if an astronaut stopped dead in the center of the voluminous workshop, where he couldn't touch anything, should there be no one to help him: would he be marooned there forever? The question proved academic, for Kerwin had found it was impossible to stop himself in the center of the workshop; once he was moving, he *couldn't* stop. Theoretically, an effectively marooned astronaut wouldn't be able to move away, though he would be able to swivel in any direction around his own center of mass. Garriott had been able to slow himself down sufficiently in the center of the workshop to experiment with this; swinging his arms and legs, he marched briskly in place, executing abrupt about-faces with a swish of his arms, as

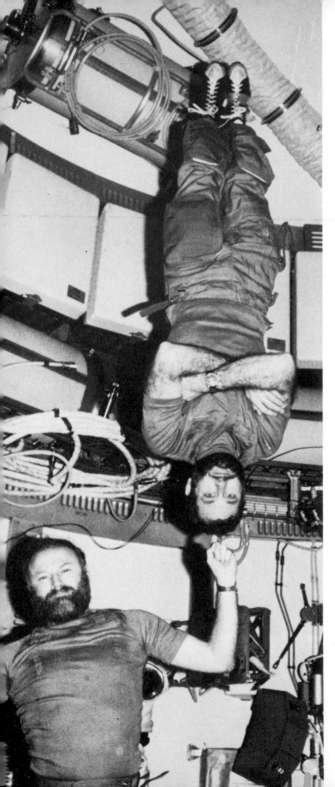

Pogue on Carr's
finger

though he were a tin soldier. Otherwise, though, he remained more or less where he was. To help the astronauts start or stop or change their speed or direction in midair, there originally had been a blue fireman's pole extending up the center of the workshop, so that the astronauts would have something to push against. As the pole interfered with their acrobatics, the men had replaced it with a long strap that also extended the length of the workshop. The strap, though, wasn't rigid enough. Once during the second mission, Bean had said, "A lot of times, you have to kind of move back and forth and get some momentum up, much like a bowstring." As this method of sort of twanging themselves around the workshop was even less satisfactory than pushing against the pole, the men had taken the strap down, too. The big volume of the upper deck was more fun when it was completely open.

The only way the astronauts could control their direction in midair was by riding one or another of two machines, called maneuvering units, which emitted little bursts of cold nitrogen gas from tiny thrusters; these sort of space motorcycles, as the astronauts thought of them, were cumbersome contraptions with quantities of knobs, buttons, dials, and spherical tanks, which made them resemble dentist's chairs. NASA had wanted Bean and the other members of the second crew to test them out in the upper deck as they might be useful if future astronauts ever had to go outside giant space stations to make repairs. The test flights were sometimes risky; one of the machines, whose thrusters were clustered at the rider's feet, was a failure because it frequently lost its balance, causing the rider to pitch forward in a complete circle, as though he were a Fourth of July fizzler gone berserk. Bean had preferred the other, whose six thrusters were spotted evenly around its frame; once he had strapped himself in, he felt as secure as if he were a tiny spaceship himself. Lousma had given a running commentary on the flight. Although Lousma's broadcasts sometimes made Skylab sound like an orbiting college radio station, he had nonetheless managed to convey better than any of the others the high spirits that sometimes came with being weightless. "OK, space fans, this is Jack," he had announced during the second mission. "The commander has left the docking station [where the maneuvering unit is berthed]. . . . There. He's at the dome lockers. Now he's rotating around, all the while six to eight inches from the dome lockers. No apparent difficulty whatsoever. Rotating slowly but surely. OK. He's passing locker number four-twenty-four. He's underneath the condensate tank now. [The tank was one of four for removing moisture from the air.] Watch head! There he is, stabilized at locker number four-thirty-two. [Bean stayed there long enough to prove he could make a repair.] Now he's backed off. Here he comes, space fans. 'Nothing to it' look on his face. . . . The kid picked a checklist ring out of midair; puts it in his jump pocket. . . . One good thing about this maneuvering unit blowing is it's helped us to locate a [checklist] card that has been missing for four days. Ooooh, he just blasted me. [Whistle.] OK. He's moving forward. Now he's moving upward. . . . OK, he's hanging on to the [ceiling] for dear life. And the kid is all snarled up in his danged umbilical. [Bean was wearing a space suit with a hose for oxygen.] Here he comes down again, space fans. Stuff blowing all over. OK, he's on his left side now. . . . Now he's completely upside down! Farewell to the workshop floor! Heads toward the dome locker. . . . Your feet are gonna hit the blue ring [of water tanks], Al! Pull them in. Attaboy, there you are. . . . Doing a beautiful job of upside down. . . . Watch your feet on the condensate tank! Attaboy. You're going to just clear it, maybe. Give yourself a little added thrust. There. That's the boy. He cleared the condensate tank handle by no more than a quarter of an inch. . . . OK, he's parallel to the workshop floor. Drifting downward, downward, downward. Pro-

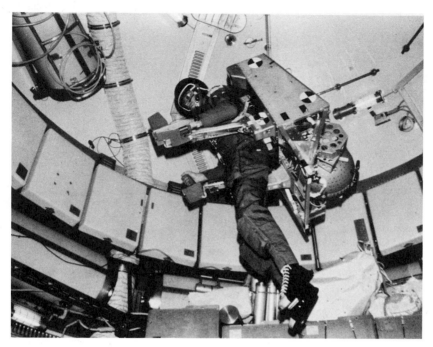

"There. He's at the dome lockers."

Bean flies the maneuvering unit sideways.

ceeding slowly but surely to the docking station, moving into the station slowly, just like the Santa Fe into Houston. Blowing a lot of gas all over. . . . Friendly observer and plane captain assisting. [Garriott and Lousma were giving Bean a hand.] He's now backed in. No. Wait a minute. Not yet. Hold that for a moment. Now, back in! You're in, you're in! OK, he's in, space fans. . . . We knew he could do it! Give him a 'well done.' "

Carr, the commander of the third crew, floated just above the grid floor of the upper deck on his stomach, like a scavenging bottom fish; looking through the triangular grid floor to the other grid floor of the lower deck was grating to his eyes. He was searching for the ultraviolet stellar astronomy experiment, a camera designed to photograph stars in wavelengths not visible on earth through the atmosphere. It was one of several instruments fastened to the floor and could be bolted hermetically to the two scientific airlocks, small square hatches on opposite sides of the second deck a couple of feet above the floor. Then the airlock hatches could be opened so that the instrument's sensor, or mirror, or other measuring device, could be pushed, or cranked, or scissored on a long boom out into space. Only those instruments intended for the airlock on the dark side could be used, as the one on the sunny side was permanently occupied by the parasol. Later, though, when the astronauts went outside themselves, they managed to attach some of the experiments requiring sunlight to the exterior of the space station.

Pogue, who was helping Carr search, floated overhead. He had to be careful not to get too far above, for in the space station's low-pressure atmosphere, he could easily float out of earshot. It was easier to talk when they were separated horizontally, for if they were on the same level, the dome overhead helped focus their voices. The floor of the upper deck was a confusing jumble of experiments because astronauts didn't always put things away properly. Bean, the commander of the second mission, had explained over television to the third crew, while they were still on the ground, "We just leave stuff around. We'll stow it before we go. Otherwise, you get in the business of stowing and unstowing if you're not careful"— an unusual philosophy for a normally shipshape Navy captain. As it happened, a lot of stuff never *had* gotten put away properly. By the time the third crew had arrived, stowage was completely out of hand. "Things were not stowed where they were supposed to be stowed!" Carr had fulminated. "We got ourselves into a mode of having to ask the ground where everything was. In some cases the ground pointed out proper places where they were stowed, and in other cases we just had to look for things till we found them. . . . Everything took two or three times as much time as we thought it would take." The mess, which had piled up during the first two missions, slowed the third crew down as much as anything else.

Leaving Carr to search for the camera, Pogue went off to look for a new mirror they had to install; it was the one that had to be cranked out of the airlock on a boom in order to reflect starlight back into the lens, which would remain inside the hatch. There were some forty thousand items stashed away in over a hundred cabinets in the space station, and Pogue bitched that none of them was ever stowed where a person might logically expect to find them. Although there were six men and a computer in Houston whose sole purpose was to help the astronauts keep track of items in the space station, the system, which had been breaking down since the beginning of the second mission because of the progressive failure of the crews to report where they put things, had now collapsed altogether. To confuse Pogue more, all the cabinets looked alike, and although they were numbered and their contents were sometimes written on the outside, the writing was small and the labels were difficult to read, particularly if Pogue approached them sideways or upside down. He had a stowage list, but he found it useless. "The stowage lists refer to numbers that are not even here!" he griped. As often as not, when he located the right cabinet, the item he was looking for wasn't in it. Pogue, who seemed to have more trouble finding things than anyone else, was reduced to ransacking cabinets at random. Several times he encountered the jack-in-the-box effect. Every time he opened a door, he felt he was looking into a pitch-dark cave, for most of the lighting in Skylab was from overhead, and it wasn't very strong anywhere. The astronauts on the first crew, who were the most inventive, had found a surgeon's headlamp in the medical-supply cabinet in the wardroom, and they had worn this like a miner's hat when they looked into cabinets; however, they had not passed on this discovery to the later crews.

After Pogue found the mirror, he returned to help Carr hunt for the camera. They passed over a big instrument that was designed to photograph the airglow, a luminous layer high above the earth that resulted from solar radiation exciting molecules in the upper atmosphere. They examined, and rejected, a squarish box that would be shoved out the airlock to collect micrometeorites. And they rejected another experiment that would have detected any gases emanating from the spacecraft, for in vacuum, many metals outgassed, or emitted small quantities of vapor, and consequently the space station was enveloped in its own atmosphere. They were getting behind the time line—the ground allotted only ten minutes to move from one experiment to another and get it set up. "You have to put away equipment, you have to debrief, and then you have to move from one position to another, and you have to look and see what's coming up, and we're just being driven to the wall!" Pogue had said during a particularly harried moment in the third mission. "There's not enough consideration given for moving from one point in the spacecraft to another and allowing for transition from one experiment to another! . . . When we're pressed bodily from one point in the spacecraft to another with no time for mental preparation, let alone getting the experiment ready, there's no way we can do a professional job! Now, I don't like being put in an incredible position where I'm taking somebody's expensive equipment and threshing about wildly with it and trying to act like a one-armed paperhanger trying to get it started in insufficient time!"

No astronauts had ever talked like this to Mission Control before. The third crew bitched and griped so much, and made so many mistakes, and fell so far behind, that the flight controllers, flight planners, and others on the ground began to speculate that possibly the astronauts were suffering from hostilities induced by isolation, as had happened in Antarctica. Or possibly the astronauts were sick. The ground wondered whether weightlessness —which changed the properties of so many other substances—might not have affected the astronauts' minds in some way. It was even suggested that the astronauts were having depressions.

As it happened, there was nothing seriously the matter with the astronauts at all; rather, the problem lay with the ground itself, which, in addition to giving the third crew too many experiments, had also started these astronauts off at too fast a pace. It took astronauts a while to learn to find things and do things in space as well as they had done them on the ground; it sometimes took three times as long to find, and set up, an experiment at the beginning of a mission as it had on earth, or as it would take at the end of a mission, when a crew had mastered the complexities of getting about and working in weightlessness. Even the second crew had taken a month before it had hit its stride; yet the ground was expecting Carr, Gibson, and Pogue to be doing the same amount of work in the middle of their second week as it had taken Bean, Garriott, and Lousma a month to progress to. The third crewmen, of course, fell behind almost immediately; while the second crew had always felt they were ahead of schedule, which gave them a good feeling, the third crew continually felt it had to hurry to catch up—a situation that was intolerable over such a long time. (In the opinion of Dr. Musgrave, the doctor-astronaut who as a capsule communicator was in frequent contact with the crews, the extreme duration of the third mission was a factor, for most people can put up with anything for the eight weeks the second mission had lasted, but twelve weeks is another matter.) The astronauts on the third crew had begun having insomnia; and to make matters worse, the ground, instead of letting up on the schedule, actually began *increasing* the work load. The proponent of this novel theory of management was Hutchinson, the lead flight director, who thought the astronauts were just being lazy and wanted to get them up to the mark. His expectations, of course, were based on what the second crew had been able to do later in its mission, but beyond that, Hutchinson himself was a big man with a deliberately intense, hard-driving air, and this may have made him unable to sympathize with the third crewmen, who not only were in difficulties, but may also have been made of different stuff. At the beginning of the second week, which was when Hutchinson began increasing the work loads, the crew's performance crumbled altogether.

It was hard to understand how the situation arose in the first place, because the director of the Skylab program, William C. Schneider, a stocky engineer normally based at the NASA headquarters in Washington, had issued instructions before the third crew went up that it was to take things easy. Schneider was not simply being humane; rather, during the long missions NASA was contemplating for the future, astronauts could not go all out the whole time but would have to learn to pace themselves. "We didn't want to repeat the experience of the second crew, which had taken its meals on the run and averaged six hours' sleep a night," Schneider said after all three missions were over. "We had told the third crew we didn't want them to get tired out or sick—that we wanted them to work a plain eight-hour day, eat three square meals, and relax a bit. Having told them that, we increased the number of experiments aboard and gave them second-mission work loads—and the peak work loads of the second mission at that! So we had given them conflicting instructions—and then we began to wonder why they were doing what we had first told them, which was to slow down." It was probably as impossible for Mission Control—or the astronauts, for that matter—to follow instructions to slow down a space flight as it would have been for the United States Marines to have followed orders to assault a beachhead gently. NASA has always placed great value on achievement, not to mention overachievement; it liked its astronauts to accomplish a great deal in order to prove how valuable men could be in space—a principle that led the third crew to conceal Pogue's illness. However far men might journey in space, they would never leave behind their earthlike ways of thinking or doing things. Corporate goals would pursue them to the stars.

"There *is* an adjustment period at the beginning of a mission, and you *do* need simple tasks then!" Carr exploded after he got back. Not only were the mechanics of getting about or finding things different in space, but so was every other aspect of doing a job—handling tools, even turning a screw. While they were in space, all the Skylab astronauts had to make a great many repairs— they fixed the broken solar panel, some storage batteries, the air-conditioning system, the guidance system, and many of the cameras for photographing the earth, the sun, and the stars. Beforehand, though, NASA had been by no means sure the astronauts would be able to do this type of work, because previous astronauts, particularly on the Gemini missions, had had difficulties performing odd jobs in weightlessness. And working inside a craft as big as Skylab, where an astronaut couldn't always reach a wall to steady himself, or where his tools might float away out of sight, posed more serious difficulties. "If you had told me ahead of time the things the astronauts would have to do to keep the mission going, I would not have believed that it could have been done," Hutchinson said afterward. The fact that the astronauts were able to do these things caused NASA to breathe more easily, for it demonstrated that, given enough time to adjust to weightlessness, astronauts could be expected to assemble huge space stations in orbit and maintain them during long missions.

Before Pogue and Carr could set up and start using the ultraviolet camera, they had to install the new mirror; the old one had become smudged and dirty. They had to think out every move ahead of time, beginning with how they anchored themselves alongside the instrument. All the astronauts, it turned out, preferred the triangular attachment on their shoes because it fitted more tightly into the triangular grid than the mushroom attachment, which slipped in and out more easily. Once the triangle was twisted into place, they didn't have to give a thought to their stability. (With the mushrooms, though, they always had to think about maintaining pressure on their feet. "You have to keep your muscles tensed, or keep a strain on the mushrooms all the time, in order to hold them in," Carr, who was always the most anxious about how he was secured, said once.) There were a great many spots in the space station where it didn't matter what an astronaut was wearing on his feet because there was no grid, and consequently he had to get along as best he could. High on the workshop walls or ceiling, an astronaut could usually find a pipe or duct to wrap his legs around, but in other places he sometimes had to jam himself in however he could. Pogue especially found this troublesome. "You have to use body English and all kinds of muscular tension in order to hold yourself in the right position to do things," he fussed once. "This means that you end up using your body against whatever pieces of hardware are available, and I have experienced numerous cuts and bruises and so forth in trying to stabilize myself while I'm working. . . . You end up using your arms and your legs an awful lot to wedge into places, and, well, sometimes it's painful." Gibson's remedy was to recommend lining the entire space station with triangular grid, walls and ceilings as well as floor—a solution, he confessed, that would make the workshop look like a birdcage.

Even though Carr and Pogue locked their feet into the workshop's grid floor, they still had a good deal of maneuverability, for they were able to reach from one end of the long camera to the other simply by swaying. Their reach—what NASA engineers called their "work envelope"—was much greater than it would have been on the ground, for they could extend in any direction so that they were almost horizontal to the floor, a position which, had they tried it on earth, would almost certainly have resulted in broken ankles. When men stand on earth, their legs hold them up; but when the astronauts were anchored in space, their legs held them *down*; as they were constantly stretching, their legs were under tension instead of compression, and some of their muscles became sore because of the unaccustomed strain. As Carr and Pogue, working on one end of the camera and then the other, waved to and fro, they looked like seaweed rooted to the ocean bottom wafted by the currents. Indeed, when they sometimes did an exercise that required anchoring themselves to the floor and then pulling up and down on a tension rope, in order to strengthen their arms, they looked exactly like spirogyra, microscopic aquatic organisms that gracefully accordion up and down.

Carr and Pogue pulled out their checklists for repairing the camera. These were in looseleaf notebooks held together by rings; Pogue's frequently snapped open so that the pages soared up in a fountain. This bothered Pogue even more than the others. "Now I understand that there was a big full-month NASA study to try to find a new kind of ring for our checklists," Pogue fumed to the B channel after such a mishap during the third mission. "Now it seems to me that a simple thing like that ought to be solved. . . . I'm right here now holding an entire book that must be about two and a half inches thick, and all the pages are loose simply because those damn rings keep coming loose all the time." Weightlessness drove Pogue to distraction. He was the most conservative and earthbound of all the astronauts, and afterward he said he would have preferred it if the space station had had some form of gravity. Pogue, of course, had suffered the most from weightlessness all along—not only did he have trouble handling things, but he complained the most about the feeling of fullness in his head, he claimed to be the clumsiest when it came to getting about, and he seemed to have the most trouble with local verticals and being upside down. Gravity, which would have cured all Pogue's ills, could have been duplicated in Skylab by rotating it so that centrifugal force would push everything against the outside wall, in which case the outside wall would have been the floor. It wouldn't require much of a spin; the first crew had made gravity of a sort one day when it ran around the ring of lockers inside the workshop, for these astronauts were pressed against the wall hard enough to provide good traction, and Pogue himself was able to do the same thing by running around the walls on his *hands*. The best-known model of a revolving space station, of course, is the one in the film *2001: A Space Odyssey,* and NASA had considered making Skylab rotate, too, but had given up the notion because it would have had certain drawbacks: it would have been more expensive, and it would have been more difficult to photograph the earth, the sun, or the stars. Most important, though, weightlessness was what NASA had wanted to study. The astronauts, too, had favored a zero-gravity space station, because, one of them had said as though he were about to climb Everest, "That's what's there." Afterward, only one Skylab astronaut—Gibson—would still feel unequivocally that way.

Pogue fumbled with a small toolbox at his belt; all the tools came out and took off in different directions, several wrenches clanging together in the lead. "Everything just comes floating out in your face, and it sounds like chimes all around you," he muttered. Whenever he carried the tools around with him, he always had what he called a gnawing fear that he would lose them; he suggested that future toolboxes have clear plastic sides so that an astronaut reaching in could pick out just the tool he wanted. There were pliers, a hammer, several wrenches, and several screwdrivers—not a large variety—and the astronauts often wished they had hacksaws, drills, and files as well. These had been purposely left out of the astronauts' tool kits because they created sawdust or shavings that would have floated about and been inhaled. They had Swiss Army knives, though, whose many blades and attachments made up for some of these lacks; best of all, there was only one tool to lose.

First Pogue and Carr had to unscrew several screws. Pogue was glad he was anchored to the grid, as once, when he had been trying to untwist a screw in a place where he hadn't been able to brace himself, he had wound up twirling across the workshop in spirals; he had recommended that there should be a knob, at least, near any screw in the space station so that an astronaut could always brace himself. Now, though, he had no such problem. Soon he had a handful of screws, nuts, washers, and other small parts from the camera; he hadn't the slightest idea of where to put them. The astronauts found that about the hardest thing they had to learn in weightlessness, and in Skylab's huge volume, was how to keep track of small items; during the second mission, Bean had estimated that he and his crewmates had lost more time looking for small odds and ends that had gotten away than in any other fashion. They couldn't put them in their pockets because they floated out. They couldn't put them down anywhere, and they couldn't simply leave them floating in the air, for however careful they were, it was impossible to position something in space without imparting some motion to it. "You have to keep coming back to it," Bean had said, after he had levitated a wrench during the second mission. "You work, and then you come back every ten seconds, you look at it, and if it starts to move, you pull it back. And then maybe ten percent of the time now it floats off and it takes you a while to find it, but everyone helps and pretty soon it turns up." And Lousma had commented after a similar experience during the same mission: "It's not difficult to handle one or two small items, but if you have many, many small items, it is difficult to handle them because they all want to float in their own different directions and all you have to do is blow on them or let them bump into each other and touch something, and they go off. You spend all your time grabbing things to keep them from getting away." Pogue stuck the screws and nuts onto a piece of tape he had stretched sticky-side-up across the camera, but this didn't work very well either.

Whenever a tool, or a screw, or a piece of the camera got away from Pogue, he had a hard time finding it again. On earth if he dropped something, he knew it would land somewhere around his feet, but in space—and inside a craft as big as Skylab—an object could take off in any direction and go any distance. Not knowing where or how far away to look confused an astronaut's depth perception; even Lousma, who was as well coordinated as any of the crewmen, had been baffled. "You're looking for something and it's right in front of your nose, but you can't see it because you're focusing your eyes further off," he had said during the second mission. And during the first mission Conrad had found that even if he were looking right at an object he had lost, he wouldn't recognize it if it were at a different angle from the one he was used to seeing it from. Designers in Houston suggested that in future space stations small objects should be brightly colored so that they would show up against the walls. Pogue and Carr weren't as bad off as they might have been, though, for any tools or camera parts they lost got caught up in the prevailing air current inside the workshop; as they went about the repair job, there was a continuous procession of screws, nuts, Swiss Army knives, and screwdrivers making their way toward the dome; and if Pogue could contain his exasperation long enough—say, about twenty minutes—he would almost always find what he was looking for on one of the ventilation screens. Sometimes he didn't. "We've lost some little weights," he complained one day when he was feeling particularly victimized by space. "I've lost some medicine. It just drifts away, and you never see it again. Some of the stuff, you lose it, and it always turns up on the screen. You lose other things, and you never see them again." Conversely, the astronauts were apt to find lots of screws and bolts and things in the air current, or on the screens, that they *weren't* looking for, and hadn't the foggiest idea where they came from. (In the event that an astronaut inhaled some small ambient object, they all knew how to perform tracheotomies.) Pogue, who perhaps thought more than the others about small improvements that might make life easier for an astronaut, suggested that cameras aboard future space stations, and all other equipment astronauts might have to repair, be made in such a way that it was impossible for hardware to float off—that latches be used instead of screws, or if screws had to be used, that they have tethers. He recommended that all walls inside any future space stations be made of pegboard—certainly handsomer from the aspect of interior decoration than Gibson's grid—so that an astronaut would always have a place to stick things; he could build up his own working space wherever he happened to be. Best of all, though, would be a workbench where astronauts could bring items needing repair, and where there would be all sorts of vises, clamps, drawers, and other attachments to hold down tools and hardware. Gibson improvised one aboard Skylab; he made what he called an "aerodynamic workbench" out of the screens in the dome, for the air moving through them was perfect for holding things down. It even sucked away loose shavings or filings so that future astronauts could have saws, files, and drills. A workbench was not much use aboard Skylab, though, because most of the equipment—the camera and some other portable experiments aside—was not movable, and it was often hard to get at, too. "I think the firing squad is in order for the people that designed that little gem!" Gibson shouted to the B channel once when he tried to reach a piece of equipment in order to fix it. "I don't want to flag that item as a major dumdum on the part of the designer, but that is pretty bad, really! That is a real hand-cutter-and-knuckle-buster operation." On another occasion, Pogue, who was beginning to think Skylab's designers had never expected any equipment to break down, actually cut his wrist trying to get at a piece of hardware in order to repair it. "I don't care if it's designed to put in ten billion hours of trouble-free operations, it may break down after ten hours," he said at the time. "And we end up replacing, modifying, patching, taping, and everything else!" Out of the third crew's bursts of anger would come, bit by

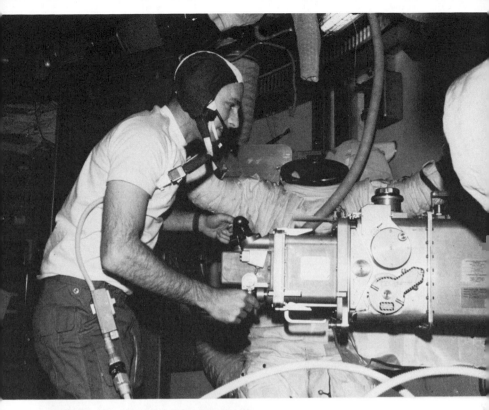

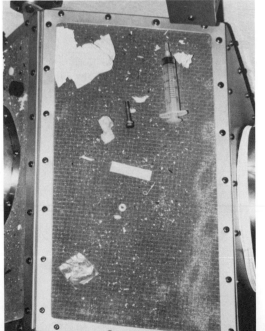

An astronaut works
with the ultraviolet
stellar astronomy
camera.

"Some of the stuff, you
lose it, and it always
turns up on the screen.
You lose other things,
and you never see them
again."

bit, a picture of what a space station, and life in one, should be like.

When the new mirror was installed, Pogue floated the big camera off the floor and began moving it toward the airlock. Though it weighed two hundred pounds on the ground, when it was in weightless state he could move it easily with one finger. Maneuvering it was difficult, though, for in spite of the fact that the camera now weighed nothing, Pogue still had to contend with its mass and momentum, something he didn't have to think about with small objects; it wobbled unsteadily until he found its center of mass and balanced it. Once it was moving, it was hard to stop; Pogue managed to halt both himself and the big camera by grabbing onto a crate that was anchored to the floor. One advantage of a large object, though, was that it didn't drift away as quickly as a small one. Pogue left the camera floating in midair when he went in search of the camera's lens and film, and he was relieved to find it was in almost the same position when he came back.

The lens and film came from the film vault, a huge box bolted to the floor of the upper deck. As the astronauts had a vast amount of photographic equipment, the vault —it was made of lead to shield its contents from radiation—had the worst jack-in-the-box problem of any cabinet in the space station. Even Gibson, who had an easier time handling things than Pogue, had had trouble when he opened it once. "That film vault, it's a god-awful mess!" he had said at the time. "I'm sitting here looking at it now, and it just kinda makes me wince. Every time I have to go over there to do something, I cringe and I say, 'Oh, God, I gotta go through this again.' . . . You open up the cover, and then you got cameras and lenses and film and tapes coming out, and it's just like the top of a snake pit, and they all start slithering out at you." Some of the film was in a row of drawers that ran down one side of the vault, and these were a special nuisance, for when Pogue opened the top drawer, lenses and film from the drawer below floated up so that the top drawer wouldn't shut; when he removed the two drawers, items floated up from the one below *them*; he had to remove all the drawers before he could get everything back together again. "And it's very frustrating," he said.

When they were all done, they were an hour behind schedule. In their butterfingered way, they had smudged the new mirror, something they managed to blame on the ground; Pogue and Carr went out of their way to make it quite plain to Mission Control that the situation would have been better if they had never been asked to do the job in the first place. "It appeared to me that the new mirror had more dust on it than the old mirror," Carr told the B channel. "And I added one smudge myself, while taking it out of its container. Whoever packed the container put the gloves [for handling the mirror] on the bottom instead of on the top. And while struggling to get the mirror out with my bare hands, it suddenly popped out, and I've just touched it right at the edge with the palm, or the heel, of my hand. The smudge is three-quarters of an inch long and extends from the edge about three-sixteenths of an inch."

His breakfast settled at last, Pogue returned to the lower deck and hopped onto the bicycle ergometer as though he were a tumbler leaping in slow motion onto a gymnasium horse. He had unzipped the lower halves of his trouser legs to make shorts, so that he looked as though he were working out in a gym. He had had trouble getting the pant legs off his shoes because of the triangle cleats, which he needed to grip the pedals. When his feet were locked in, he grabbed a white hose that was snaking wildly in the air beside him and clamped its mouthpiece between his teeth; he also attached a pressure garter to his arm. Then he started pedaling. The air he exhaled through the hose was automatically analyzed to see how much oxygen he was absorbing, and this figure was plotted against the force of his pedaling, and also against his pulse; the more oxygen he needed for a given amount of pedaling, and the lower his heart rate, the better condition he was in. In space, where there was no gravity for Pogue to work against, and where getting about or lifting things placed no physical strain on his body, his unused muscles were apt to deteriorate; his legs were particularly vulnerable as they were never used as they were on earth. His bones, particularly his leg bones, which no longer had to support his weight, were losing calcium. Weightlessness was changing the astronauts' bodies as surely as it altered the shape of water. The flight surgeons liked to think that all these trends might be helpful adaptations to space, on the theory that in weightlessness, where an astronaut didn't need so much of everything—not only muscle and calcium, but electrolytes and red cells as well—he might be better off without them.

All adaptations, of course, were not necessarily beneficial. Pogue's heart was no more immune from weakening than any other muscle in his body; indeed, his heart was particularly apt to atrophy as it no longer had to pump blood against gravity; for this and other reasons, it decreased in size. The blood vessels in his legs weakened, too, for they no longer had to exert a pressure against the blood which on earth normally pooled there; in space, where this didn't happen, the veins in an astronaut's lower body were apt to lose their tone, creating difficulties when he returned to earth. (The astronauts' legs were such a nuisance in space, where they not only caused cardiovascular and muscular problems but were also something of a hindrance to an astronaut's mobility, that later one doctor at the Skylab Life Science Symposium, Dr. Robert P. Heaney of Creighton University in Omaha, Nebraska, quite seriously proposed that amputees be considered for future space missions. "We have seen that [the legs] are nearly useless," he said. "Their tissue requires food and consumes oxygen, and if we exercise them, they consume even more. The ultimate fuel cost of legs on long missions must be really staggering.") For better or for worse, the Skylab astronauts were still very much attached to their legs; and on mornings when they didn't do the bicycling, one or another of them hopped into the big aluminum barrel in the experiments room, the lower body negative pressure experiment, whose purpose was to simulate the effect of gravity upon an astronaut's cardiovascular system, particularly in his lower body. On the thirteenth day of the first mission, Kerwin had jumped into the LBNP; a black rubber skirt had been fastened around his waist to make a tight seal between him and the barrel. Then the air inside had been vented into space; the vacuum drew his blood back toward his feet, the way gravity would. Kerwin had not enjoyed being inside the LBNP, particularly as he could feel quite faint if too much blood went to his legs; the weakened blood vessels ballooned so that Kerwin had been unable to tolerate the maximum suction. "Terminated the run early, based on a feeling of a little dizziness and cold sweat," he had reported afterward. The most disconcerting part of it, he had confessed, was a feeling he had had that he, too, might be sucked out of the can, along with the air, and spewed out into space.

The astronauts preferred the bicycle ergometer. Cycling was harder in space than on the ground, for they didn't have their weight to help them. To keep himself on the bike, Pogue put a pillow on his head and braced himself against a pipe on the ceiling. He gripped the handlebars tightly. Because the triangles on his shoes were locked to the pedals, he could pull up on one pedal while he pushed down on the other, and this also helped keep him in place. While he was working out, the experiments compartment couldn't be used for anything else; hoses and legs flapped about so that anyone else in

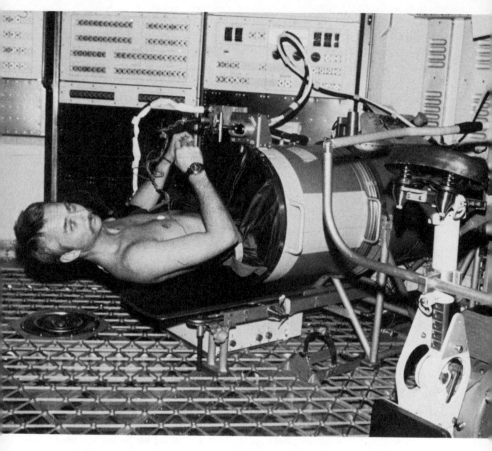

Garriott lies in the barrel of the lower-body negative-pressure experiment. Earlier, when Kerwin tried this, he feared that he might be sucked out of the can and spewed into space.

the room had to stand clear. Any bikes aboard future space stations should have special rooms of their own, Pogue said. It was hot work, and as the heat didn't rise, but instead lingered suffocatingly about him, he was grateful for a fan the second crew had mounted on the wall nearby. He wished he had more changes of clothes—the fireproof cloth they were made of, he said, was not good for athletics. "It stinks, is what it does, after a couple of days' use," he bitched. The later crews had their daily cycling divided into two periods, and this added to their problems. "We would prefer not to have our exercise sessions split up into two sessions," Carr said once. "Now the reason for this is that you do some heavy exercise and then you've got to clean yourself up before you can do something else. And if you have two exercise periods, all that means is that you've got two [cleanup] periods. . . . And don't take my fifteen-minute [total] cleanup time and divide it into two seven-and-one-half-minute cleanup times, because that isn't gonna hack it either. . . . On an average period of exercise, it's gonna take you twenty-five minutes to clean yourselves up afterward. And if you don't believe it takes that long, go over to the Skylab [simulator at the Space Center], take a washcloth, and try to take a bath. Wash your entire body with nothing but a washcloth and a pail of water. And see how long it takes you." As the third mission went on, Carr, as befitted the commander, concentrated his remarks more and more to the kind of schedule Mission Control was giving them and the kind of life Mission Control was expecting them to lead, leaving the more practical matters of equipment design for Gibson and Pogue. Carr would later regret that he hadn't been as blunt as often as the other two.

Hot work as the ergometer was, it missed a great many muscles in the astronauts' bodies, such as the arm muscles or the big strap muscles across their backs. Sometimes the astronauts pedaled with their arms, though not often enough to do them much good. Even in their legs, though, there were a number of muscles the ergometer missed—in particular, their calves became flabby, and this would later give them trouble walking when they returned to earth. To give his calves more of a workout, Gibson used to go around with his shoelaces loose and his triangles well forward, so that he would be on tiptoes more, but this didn't do much good. What did do some good, though, was a new item of equipment jammed aboard the last mission, a treadmill. Carr tried it one day—it was simply a slippery sheet of Teflon that he bolted to the grid floor just behind the bicycle; then, anchoring himself to the floor with a harness, and balancing himself with a hand on the bicycle seat, he simply walked along, his feet slithering on the slippery Teflon. He wore a pair of socks so that his feet would slip along more easily. Occasionally he bruised a foot on one of the bolts that held the Teflon down. It was hard work—Carr found that five minutes at a time was about all he could handle—but, he added, "It really gives the tendons and everything, and the tops of your feet and your calf muscles, a real workout."

The astronauts still preferred the bicycle ergometer. Increasingly they rode it simply to keep themselves in shape, and they found it made them feel better, too. "I got to where I wanted to get on that bicycle, especially if I'd [spent] two or three [hours at the solar-telescope console] where I just sat in zero gravity and floated," Conrad had said. "Your body tells you to do some work, pump some blood, it'll make you feel better." After a good workout, Kerwin had told Mission Control that he felt "that strong glow of health" he was used to after similar workouts on the ground. It made the men look better, too, for the extra flow of blood to their leg muscles reduced the puffiness of their faces. It got so that the astronauts complained bitterly if their time on the ergometer was cut into by the flight planners. They never felt they were getting enough exercise; Conrad had once estimated that an astronaut would have to work out a total of *five hours* a day simply to equal the effort of just being, and moving, on earth. Conrad had been so anxious to keep in shape that he once had pedaled for ninety minutes straight, which, as it was also about the duration of one Skylab orbit, led him to claim that he had bicycled all around the world. The astronauts on the second mission had spent more time on the ergometer, and the third crew exercised most of all. They found that they pedaled better if they had music to listen to, and they thought the pedaling would have improved if the ergometer had been near a window so that they would have had something to watch while they rode. All this time and effort expended on keeping in shape caused Dr. Heaney to wonder at the Skylab Life Science Symposium whether it wouldn't be better to send into space men who were *out* of shape who wouldn't worry about becoming flabby. "I have found myself asking, repeatedly, why there is this quite extraordinary emphasis on physical fitness for function in a weightless environment," he said. "Great muscular strength and endurance, which have obvious survival value in the jungle, are all but redundant in a zero-gravity environment. . . . We can select individuals [for future space travel] already adapted to something closer to zero gravity. Here, I refer to sedentary, skinny, small individuals, like myself, who would be better suited than these athletes."

As the astronauts had a marked disinclination to wither away, they repeatedly asked the ground how they were doing; they kept their records on the wall of the experiments room and consulted them as avidly as hospital charts. Their performance on the ergometer remained remarkably steady; after an initial slump that was probably due simply to learning how to ride a bicycle in space, they quickly returned almost to their preflight levels, and after that, as one doctor who kept track of their daily performance put it, "You could lay a ruler across the

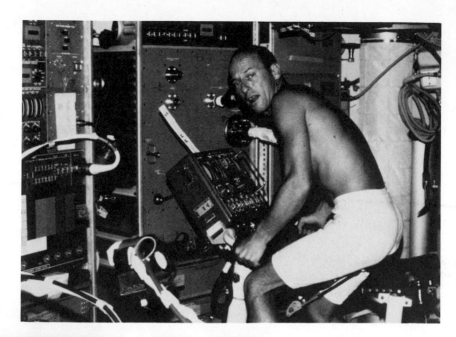

Conrad, the commander of the first crew, rides the bicycle ergometer. Once, when he pedaled for ninety minutes straight, the duration of one Skylab orbit, he claimed that he had bicycled around the world.

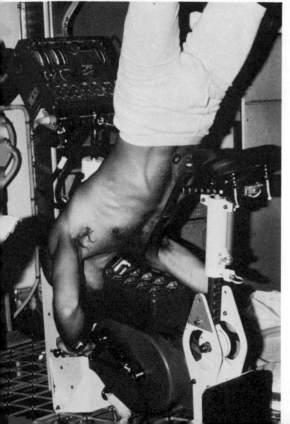

Conrad, legs in the air, bicycles with his arms to exercise his upper body.

results." In contrast, their performance on the LBNP continued steadily downward until, between the twentieth and thirtieth days of the flights, most of the astronauts found they were unable to complete their runs in the barrel, though Kerwin, who turned out to be the worst off, had failed on the thirteenth day. An odd thing happened, though: those crews that stayed up longer found that their performance in the LBNP, as well as the condition of their cardiovascular systems, began to *improve* between the thirtieth and fortieth days; after that, it stabilized at an apparently safe level, and in the case of some astronauts, their performance continued to improve almost to preflight levels. They stopped losing muscle tissue around the fortieth day, too, for that was when they stopped losing weight; after that, some of the astronauts even began to gain weight. The third crew, which was up the longest, lost the least. All in all, the flight surgeons were in for a pleasant surprise. Most of the other trends they had worried about leveled off even earlier, though the flight surgeons did not know this until after they had examined the blood and other samples the astronauts brought back with them. The loss of fluids stopped between the tenth and fourteenth days, and with it most of the other trends associated with the fluid loss, such as the depletion of electrolytes; as a result, the astronauts suffered no heart arrhythmias. The loss of red blood cells leveled off then, too, and the astronauts on the third crew even began to generate new red cells while they were still in space.

The flight surgeons were baffled, for they had expected that the longer a man was weightless, the worse his condition would be. Clearly, exercising on the ergometer was a great help—the healthiest astronauts were invariably the ones who had done the most daily cycling—but the flight surgeons didn't believe it accounted for all the improvements they were seeing. For example, it was hard to see how it caused the improvements on the LBNP or the regeneration of red blood cells observed in the third crew. Clearly something else was going on that the flight surgeons didn't understand. And they worried, too, that the astronauts were adapting to weightlessness too well. They knew a penalty would have to be paid when the astronauts returned to earth, and the greater the adaptation to weightlessness, they suspected, the higher the price would be.

While Carr and Pogue had been cycling on the ergometer and repairing the ultraviolet camera, Gibson, the science pilot on the third crew, had gone straight up through the dome to the docking adapter to start observing the sun. The docking adapter was a long tunnel of a room with consoles, instruments, and boxes jutting radially from its cylindrical walls. Even Gibson, who could keep his bearings better than most of the other astronauts, was momentarily confused, because he had left behind the workshop's strong architectural sense of up and down, or local vertical. If the workshop was like a can set on end, the docking adapter was like a can lying on its side; without gravity, and with all the fixtures protruding every which way from the walls, there was no suggestion of which way was up. To make matters worse, the command module—visible through the hatch at the far end of the docking adapter—had a strong vertical of its own, but this was the opposite of the one Gibson had just left, for the command module's floor and ceiling were upside down in relation to the workshop's. Consequently, any astronaut traveling the length of Skylab, from workshop through docking adapter to command module, had a tough time of it, for he had to deal with all three types of local verticals. After Carr had done this once, he complained on the B channel, "I get, you know, [one local vertical] embedded in my mind, and I whistle [out of the workshop] through the docking adapter and into the command module, and zingy! All of a sudden it's upside down. . . ." He felt it would be dangerous if an astronaut, in an emergency, had to work hastily between compartments; he might throw a switch the wrong way. Carr, and most of the other astronauts, would have liked the space station better if everything had been the same way up—though few of them would go as far as Pogue, who wanted gravity.

If the astronauts had a hard time orienting themselves inside a place like the workshop, which had a single, well-defined local vertical, it was almost impossible to do so in the docking adapter. Among the consoles, instruments, and boxes jutting radially from all around the cylinder, the local vertical was anywhere around the walls that an astronaut happened to be working. The docking adapter had been built that way because the designers of Skylab had wanted to see whether men could get along without a single vertical; if the astronauts liked the docking adapter, the designers had thought, then they could use the entire volume of a room in planning future space stations; they could, for example, multiply the use of a room sixfold by putting equipment not only on the floor but on the ceiling and the four walls as well. The astronauts hated the docking adapter. Pogue, who had the greatest trouble adjusting to new verticals, and who therefore regarded the docking adapter as his own personal *bête noire,* exploded, "Well, all I gotta say is, if you want a very good example of how not to design and arrange a compartment, the docking adapter is the best example. Boy, it's so lousy, I don't even want to talk about it until I get back down to the ground, because every time I think about how stupid the layout is in there, I get all upset. . . . If you want an example of how not to lay something out, there it is. Boy, go in there and take a good look, because that's the way you don't want to do it."

By and large, most of the astronauts turned out to be so reluctant to give up the idea of a single vertical, such

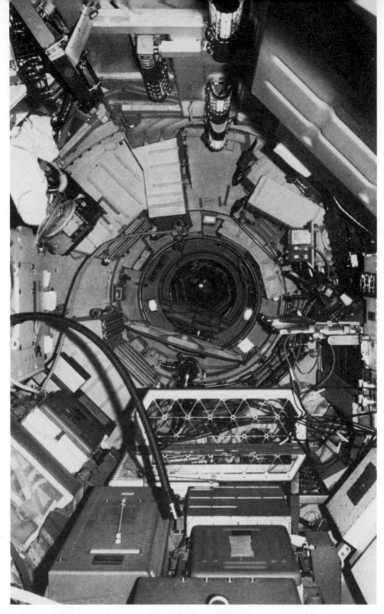

The docking adapter is a long tunnel of a room with consoles, instruments, and boxes jutting radially from its cylindrical walls; with no clear vertical, the astronauts had a hard time orienting themselves inside it. Pogue, who had the greatest trouble, said, "If you want an example of how not to lay something out, there it is." The view through the hatch extends all the way along the main axis, to the trash airlock in the center of the lower deck, some ninety feet away.

as they had enjoyed on earth, that the designers despaired of their more freewheeling plans. Bean, who, like Pogue, was among the less venturesome of the astronauts when it came to these matters, said, "There's been some thought about mounting some furniture on the floor, some on the walls, some on the ceiling, but this doesn't work out. You tend to orient yourself when you're in a room, even though you're in zero gravity, and when you orient yourself, you should find everything is the same. You don't like something up, something under. . . . You like things to be orderly like they always are on earth. Now, if you want to put everything on the ceiling instead of the floor, we can sure handle that. It's just that we don't want half and half." Like compass needles, the astronauts seemed to require a single strong force in a room that they could lock their own personal verticals onto. Almost all the astronauts, who had ranked the earthlike lower deck as the most successful part of the space station, went on to declare the docking adapter the least successful; and they never became more reconciled to it, no matter how long they remained in space.

There was one major exception among the astronauts, and that was Gibson. If Pogue was the most earthbound, Gibson was the best adapted to life in space—at least he was the most able to break away from the patterns of life on earth. And as he was also one of the most intelligent of the astronauts, as well as one of the most perverse, his dissent may have been important. Alone among the nine Skylab astronauts, Gibson actually preferred the docking adapter, with its lack of any single vertical and its kaleidoscopic radial arrangement of boxes and consoles. "Some people, I guess, do knock the docking adapter, but I kind of like having the walls as the working space," he said once. "I'd say the best part of the spacecraft is the docking adapter." And the workshop, which everyone else liked because of its strong local vertical, got a low mark from Gibson, who said its earthlike arrangement wasted a lot of wall space. Gibson, along with Kerwin, thought more imaginatively about the strange perspectives in space. The two science pilots complemented each other very well, for Kerwin, a physician, tended to think more subjectively about what it felt like to be inside a space station; while Gibson, a physicist, was more objective, thinking about the space station itself and how it could be improved. He was forever launching new models for space stations over the B channel. He wanted to build one that would be a giant version of the docking adapter—one that was the size, say, of the workshop—with consoles, experiments, and work stations all radiating from the walls. Perhaps Gibson had a stronger sense than the other astronauts of his own personal vertical, so that he was less adrift in a room without one; or perhaps his sense of balance in his inner ears had been less affected by the upward migration of fluid. Whatever the case, NASA would do well to find out, with an eye to selecting future astronauts.

Though it was his favorite place, Gibson still had plenty of trouble with the docking adapter. "It is one of the biggest mysteries in the world when you go in there to find something," he griped. Before he floated into it, he had to think for a minute where on the round wall the console for the solar telescope was, and then he twisted himself in midair so that he would be facing it when he arrived. Everything was screwed around in relation to everything else, and it always bothered a man using the solar console—at the far end from the hatch by which Gibson had entered—to glance back at the other big console in the room, the one for the earth resources experiments, which was near the hatch from the workshop, for it was partway around the wall in relation to himself. The boxes, instruments, and consoles swirling around the white walls were a maze of black, blue, and gray. To make matters worse, the numbering system in the docking adapter, where it should have been the clearest, was most chaotic, and even Gibson said, "The guy, when he did that, it looks as though he just kind of flipped numbers in the air and scattered them all around, and whatever way they came out, that's the way it was." The illogical numbering completed the looking-glass quality of the room. Pogue, who resembled Gibson only in his use of invective, reserved some of his most bitter words for the final confusion the numbers created. "Locatability [here] is so bad it almost looks like you had to go out of your way to design it that way!" he said during the third mission. "I mean, to try and find a [cabinet] in this place! Sometimes the numbers are there, but they're hidden; you can't find them. If you know where [something] is, then dang it, you don't even need the number; and if you don't know where it is, the number doesn't do you any good." Though Pogue was the most down-to-earth, not to say earthiest, of all the astronauts, with little of the imagination of either Gibson or Kerwin, he would nonetheless be the one to understand best why he, Gibson, and Carr were so angry so much of the time, and to see most clearly what had gone wrong with the third mission.

As Gibson made his way along the docking adapter toward the solar console, he had to be careful not to bang into any of the boxes that jutted toward him like the knobs and spikes inside an iron maiden. Lousma had once likened the docking adapter to a boiler room where a man had to be on the lookout for what he called "head knockers." As he floated along, he steadied himself by gripping onto handrails, which had been painted blue so that an astronaut wouldn't grip something else by mistake. He constantly worried that he would grab onto or bump into some important piece of equipment. As the console for the solar instruments was also the main console for running the entire space station, a misplaced hand or foot could turn out all the lights in the space station or put Skylab into a new orbit. "One thing I have a deathly fear of is grabbing those rate gyros and setting us off on a wild goose chase," Gibson fretted once. (The rate gyros, which the third crew had had to replace, were the small electronic gyroscopes that were the essential component of the craft's automatic guidance system; they rattled and banged so that no astronaut was ever able to catch a nap at the solar console.) Gibson later told Mission Control that aboard the future space station he was fond of constructing there should be plenty of corridors and hallways, and plenty of small rooms opening off of them, so that no console ever had to be in a passageway.

Gibson thought more like an architect than anyone else aboard Skylab, and possibly more than anyone connected with the program on the ground. The first house in space might have profited from the services of a professional architect, yet none had had a hand in planning Skylab, except perhaps in the very earliest stages of its design. Skylab was the product of engineers, who were not as likely as architects to consider man's relationship to the structure they were making. One architect who had an early connection with the program had felt that the engineers almost forgot that men would be living in Skylab, for in his opinion they seemed to expect the astronauts to get along as best they could among the consoles and black boxes—just as other engineers who were planning the astronauts' daily schedule expected them to live through days that, when plotted out on paper, looked like the diagrams for a complex piece of circuitry. One of the flight surgeons, Dr. Charles E. Ross, was convinced that not only overplanning the astronauts' days, but also awkward design of the space station itself, played a part in the irritation of the third crew. "From the point of view of habitability, Skylab was really thrown together," he said. Good architecture, he felt, would have had a demonstrably soothing effect on the astronauts.

There was a cassette tape recorder tethered to a string above the console, and the first thing Gibson did when he got there was to turn it on. It was easy to tell who was manning the console by the music, which could frequently be heard over the radio in Houston; if it was classical music, the sun watcher was most likely one of the science pilots, particularly Gibson or Kerwin; but if it was country music, the watcher was one of the others, most likely Bean. "The music just really passes the time," Bean had said once on the B channel, which reverberated with the twanging of an electric guitar. "The solar console is the only time you really have sort of by yourself. You come up here and spend two or three hours, and it's really pleasant work." This was one of the few times any astronaut called anything in the space station "pleasant," and when one did use that word, he was invariably a member of the second crew. One reason Bean and the other astronauts liked sun watching was because it was about the only time during the day when they had any privacy. The docking adapter was so isolated that Gibson couldn't hear Pogue and Carr down in the workshop; if he wanted to talk to them, he had to use the squawk box or shout through the hatch. As people do with hideaways, the astronauts had turned the console into a homey place; overhead was a bulletin board where they had stuck all manner of items—memos, shoulder patches with emblems of the Skylab missions, and Polaroid photos of the sun.

Gibson anchored himself to a small patch of grid that served as a floor in front of the solar console—two long, straight panels, one above the other, which fitted into the curvature of the docking adapter wall. Gibson, who spent the most time watching the sun, wished the console, instead of being flat, curled protectively around him so that he wouldn't be kicked in the head every time someone passed behind him. There was no chair, but as Gibson was standing in the slight crouch of the neutral G position, he looked as if he were sitting in an invisible one. Once there had been a real chair, with a conventional seat and back, but the astronauts found they could do better without it. The first crew had removed it and no one had ever put it back—thereby effectively disposing of any questions about the utility of chairs in space. Because the console had been designed for a seated astronaut, though, Gibson found it too low; his stomach muscles ached from the continual bending. He reached up to press some switches in order to jockey the space station into the proper position for viewing the sun. The attitude controls, and the other switches for operating Skylab—ones for heat, light, pressure, and so forth—were at the top of the console or far out at the sides, but there weren't nearly as many of them as there were on the dashboard of the command module, because Skylab was supposed to be a house, not a vehicle, and most of the routine maintenance was looked after on the ground so that the astronauts would be free for other work. Gibson turned Skylab so that the tall telescope tower on top of the docking adapter, which housed the six solar cameras, pointed at the sun—a position called the solar-inertial attitude. Three huge wheels called momentum-control gyros, which had been mounted at right angles to each other inside the telescope tower and which rotated nine thousand times a minute, kept the space station pointing the right way through their spinning; Gibson adjusted Skylab's attitude by adjusting the speeds of the different wheels. During the third mission, when one of the wheels broke down, and a second seemed about to fail, the astronauts considered making the station rotate in order to stabilize it so that it wouldn't tumble; if they had done this, they would have inadvertently turned Skylab into an artificial-gravity space station.

The controls for the six cameras, which Gibson would be using more often, were nearer to hand, at the center of the console. He pushed switches to open sliding doors in front of the cameras, which were set into the top of the telescope tower straddling the docking adapter outside; the doors protected them from the cold during the nighttime passes behind the earth. Since Skylab's orbit was stationary in relation to the stars, it spent greater or lesser amounts of time in sunlight as the earth moved around the sun. There was one period when Skylab was in sunlight almost continuously and the doors could safely be left open. Because of these differences, there were what amounted to seasonal temperature variations inside the space station. The docking adapter, which in addition to its other problems was the coolest part of the space station, could become almost unbearably chilly during those wintry periods, when Skylab spent greater amounts of each orbit in the shade. As Gibson expected to be at the console for several hours, he pulled on a pair of gloves; later he suggested that any future space station be placed in an orbit such that it would be in sunlight all the time.

Directly in front of him, at the center of the console, there were the two round television screens for viewing the sun. Gibson's main purpose in becoming an astronaut seven years before had been to see the sun from above the earth's atmosphere. Having received his doctorate in physics from the California Institute of Technology, he had written a book about the sun; it is the only book ever written by an astronaut on a subject other than being an astronaut. (Less than a year after he returned to earth, Gibson would leave NASA and join a private research firm in order to devote his time to the analysis of the pictures he and the other Skylab astronauts had taken of the sun.) From the moment he first used Skylab's solar telescopes, Gibson felt that a solar physicist couldn't really say he had seen the sun until he had done so from space; the view from below the earth's atmo-

Garriott "sits" at the solar console, though there is no chair beneath him, the one provided having been removed by the first crew. As the console was too low, the astronauts' stomach muscles ached from the continual bending.

Gibson sitting at the solar console, right side up . . .

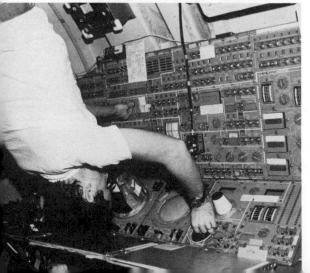

. . . and upside down. The main controls for the space station itself are on the upper panel.

sphere, which filtered out most of its wavelengths, was pale by comparison.

From Skylab, the sun was a disk about the size of a dime held at arm's length that blazed so incandescently that no astronaut could look directly at it. The six cameras in the telescope tower photographed the sun in five different wavelengths—all invisible on the ground—and as each wavelength was generated by a different temperature, and as the different temperatures were generated in different interior zones of the sun, the cameras collectively provided a cross section through the sun's upper layers. Two of the cameras were sensitive to the H-alpha wavelength, which was considered the best for overall monitoring of the sun, and the image from one of these instruments constantly occupied one of the two screens on the console; it had a cross hair for aiming. Through it, the sun was a roiling sphere of white gas that was so full of black specks churning slowly that Carr once likened it to a "great big bowl of oatmeal with pepper on it." Gibson could bring up the image from one or another of the other four instruments on the other television screen. Through one camera, the X-ray telescope, which photographed the regions of greatest activity and heat, the sun was a medley of orange and black like a Halloween jack-o'-lantern; the most active regions showed up as orange spots against a blue background. Through another, the Extreme Ultra Violet Monitor, it was a disk of splotchy colors that looked like a chart for detecting color blindness. Gibson was fascinated; the longer he sat at the console, the more ambitious and solar-oriented his future space station became. Before he left Skylab, he suggested there should be enough solar physicists aboard to monitor the sun in eight-hour shifts around the clock, and that two or three should be on each shift.

Gibson and the other astronauts clearly preferred watching the sun to anything else they did aboard the space station, excepting possibly for watching the earth, which they would be doing that afternoon. "You have something interesting in front of you in the way of displays, and you've got a lot of high-powered observing instruments available, and you're challenged to make the best of the situation," Gibson told the B channel. "I find that so many of the things we have on board, you do by rote, or by checklist, that you don't think about them; they're just push-the-button-and-make-sure-it-works type of experiments. There's nothing wrong with some of those; you can learn an awful lot from them. But they sure are hard on the operator if you're going to do that all day. And I think it is the solar telescope, and the out-the-window observations of the earth below, that keep us challenged and mentally awake. Without these, you'd be ready for the rubber room when they brought you back." More than any of the other astronauts, Gibson resented the standardized, mechanical routine imposed on them by Mission Control; when he was watching the sun, though, he felt he had a measure of independence. He felt he worked better this way, and as a direct result of his sun watching, he recommended that future astronauts should be scheduled loosely all day long, no matter what they were doing. "I think in the future the ground should give the astronauts the bare framework of a schedule, together with a sort of shopping list of things for them to do, and then let the guys on board figure out the best way of doing them," he said. This was not altogether to the liking of Mission Control, however, and Hutchinson, the lead flight director, insisted that Skylab was far too complex for astronauts to operate by themselves, Gibson's taste of freedom on the solar console notwithstanding. "So many jobs interfere with one another!" Hutchinson said after the third crew had returned to earth. "What if a guy gets an instrument focused on a star, and just then his buddies in the docking adapter maneuver the vehicle around to look at the earth? Or

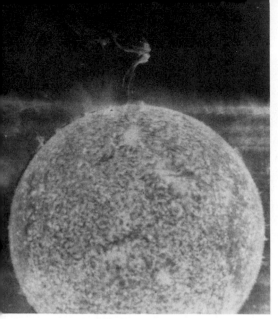

The sun photographed in extreme ultraviolet light. The plume of material erupting at the top is a half-million miles long; this picture proved for the first time that such eruptions, of helium, could stay together at that altitude.

The sun seen through the H-alpha instrument the astronauts used for monitoring and aiming; the cross hairs are lined up on a solar flare. Through this camera, the sun looked like "a great big bowl of oatmeal with pepper on it."

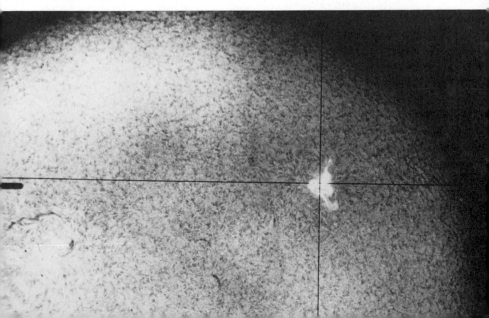

what if a guy starts riding the bicycle ergometer, jiggling the space station, while another guy is taking a long film of a solar flare? Now, say that I gave the crew a rough framework of a schedule that said, for example, 'Do five orbits of solar work followed by two orbits of earth resources passes over Africa.' They might get so super-interested in the sun that they didn't get ready in time for the earth resources passes and missed an important target on the ground! With so many constraints, I'd say they're bound to screw something up!" Hutchinson clearly would have been happier if Skylab had been a mechanical, unmanned satellite; indeed, the ground frequently acted as though the astronauts were simply components in some elaborate electronic machine operated solely by Mission Control.

Sun watching made the astronauts feel less mechanical, for their observations were their own in a way that little else was that they did aboard the space station. There was always something for Gibson to look at. Light smudges called active regions appeared and moved slowly across the edge of the sun's disk, disappearing finally around the back side. Streams of material called filaments shot high into the corona and disappeared. Flares leaped up and dropped back in great arcs. Prominences rose in bumps, only to burst later as bubbles; one bubble, almost as big as the sun itself, emerged while astronauts were watching and burst out into the corona, showering electrons onto the space station and the earth. Many of the events moved from one wavelength to the next as they traveled vertically up through the sun, from layer to layer. Gibson wished he had more television screens so he could keep track of all the layers at once. On most of the wavelengths, pinpoints of light, each as big as the earth, kept appearing and disappearing, five hundred at a time, and none lasted more than eight hours. Other features called supergranulation cells were found on many layers, too. For a long time, solar physicists had wondered how the sun managed to radiate so much heat, for its surface was relatively cool—only about five thousand degrees, compared to millions of degrees in the sun's interior and to hundreds of thousands of degrees for the corona, or atmosphere, farther out. Now it looked as if the pinpoints and the supergranulation cells, which up until then the scientists had thought existed only on one or two layers, were really chimneys of a sort, through which the sun's heat, and other forces such as its magnetic field and the solar wind (a stream of electrons radiating continually from the sun), passed from deep inside the sun on out to space without heating the sun's surface.

As Gibson hunched over the console, what he was particularly on the watch for were flares, those bursts of energy that flung great arms of flaming gas into space until the sun's gravity made them arch back again. Sometimes there were several at the same time, or in sequence, like slow, majestic lightning flashes. Whenever a flare occurred, the astronaut on duty would press buttons to make all the cameras go at once. An alarm was rigged to a radiation wheel to ring whenever there was a flare and no one was watching. There were frequent false alarms, for there was a rent in the earth's magnetosphere that protected the earth from the sun's radiation, and whenever Skylab passed under this spot, high over the South Atlantic, the flare alarm went off. The alarm kept the astronauts hopping. "Just a minute!" Lousma had shouted irritably to the ground once. "I gotta get a bit of my beef hash here. A little late breakfast this morning because of that flare." The extra radiation caused them to see a shower of flashes like blue-green pollywogs as electrons impacted their eyes. They dutifully reported them to the ground. "MARK! Left lower center. And this was a curlicue," Pogue said, as though he were spotting planes during an air raid. "It was about, oh, like a quarter of an inch long. It seemed to go down and curl around.

It's the first curved streak I've ever seen. . . . MARK! Both eyes. MARK! MARK! Streaks going in different [directions]. MARK! Golly, they're really zapping me. MARK! MARK! MARK! MARK! MARK! MARK! I'm getting zapped by the tadpoles. MARK! MARK! MARK! Oh, man, I don't know what happened. All heck broke loose."

Sun watching was complicated work, and Bean, who stood a little in awe of scientific things anyway, had said, after a long tour of duty on the console, "It makes you a little humble, because there's no way to go very long on any of this without making a mistake, and you just hope you don't make any that are too large. I don't think we have made any large ones yet, but I'll tell you, you're doing something every minute. You're throwing a switch, or reading something. You think you are good till you get here, and then you find out that you're not as good as you think." While the cameras were running, the astronaut on duty had to make sure that Skylab was holding very steady, for in order to line up the slitlike aperture of one of the instruments along the edge of a sunspot, the entire space station had to hold as still as if it were zeroing in on a dime a mile away. Sometimes, though, the cross hairs on the H-alpha monitor would wobble, and of course the photography would be blurred. On these occasions, Gibson floated to the hatch and shouted down to the two others in the workshop to stop bouncing around. Even though the astronauts weighed nothing, their mass was sufficient so that every time they caromed off a wall downstairs, it sent a slight quiver through the entire hundred-ton space station. When the first crew of astronauts had run around the water tanks, they had ruined a time exposure. A cough exerted the equivalent of fifteen or twenty pounds of force, and even pressing the button on the water fountain in the ward-room could make the cross hairs jump across the sun.

Clearly, Gibson's future space station was going to need bigger and more sensitive momentum-control gyros to hold it steady. In the meantime, though, the scientists on the ground, who fretted that their pictures were being ruined, could do little more than send messages to the astronauts to cool it. The scientists, who to a limited extent could operate the cameras from the ground, looked forward to the periods when the astronauts were asleep, or the periods between missions when Skylab was empty, for any pictures they took then were bound to be unblurred.

Indeed, a number of scientists—not unlike some of the engineers—had wondered beforehand whether they wouldn't do better using unmanned spacecraft for their pictures ̓rather than relying on the quirky astronauts; however, most of them changed their minds when they saw the film the astronauts brought back. When the scientists ran the film from the H-alpha cameras through a projector at high speed in order to get a sense of the movement of the sun's disk, they were delighted at the way every time anything interesting occurred on the sun —a prominence, an active region, a flare—the astronauts had focused on it. "Zip! There went the cross hairs right to it," one scientist said, "and there was a record that was terribly impressive of how the astronauts were able to take advantage of every target of opportunity." During the missions, the ground was continually advising the astronauts about what they should be focusing on, and the astronauts, for their part, found the scientists in Houston indispensable. "If I didn't have them, I'd be lost," Gibson admitted. "I'd be operating on information two months old, and I don't think I'd be as stimulated." Gibson talked directly with the solar physicists in Houston only twice during his three months in space, though, and even then these were not private conversations where he could talk freely, for all the flight controllers were listening in. As the talk was quite technical, Gibson felt distracted because he could almost *see* "an awful lot of eyeballs" rolling into the backs of an awful lot of heads down in Mission Control. The astronauts aboard Gibson's own space station would be allowed to talk with scientists on the ground whenever they wanted, over a private hookup that would circumvent the flight controllers— something that Gibson was in favor of doing in many areas.

Normally, the astronauts informed the ground about what the sun was doing in lengthy monologues delivered over the B channel. Gibson and Garriott, who had backgrounds in solar physics, were naturally quite scientific, but Kerwin, the science pilot of the first crew, had seemed able to get across what he was seeing in the plainest English. Of all the astronauts, he was always the most literate. "Friendly B channel, this is the science pilot," Kerwin had said one day. "I'm going to amaze the solar scientists with an example of erudite misuse of words. I have done a prominence survey around the limb of the sun. And in addition to seeing the old prominence number sixty-two just disappearing behind the west limb, I have on the east limb a beautiful-looking prominence, which we call a bifurcating *tree-trunk* prominence, which we report in hopes that it may portend further activities coming up around the east limb. . . . Your science pilot at 16:10 with some more solar notes. . . . The prominence previously observed to be departing from the sun in the vicinity of the northeast limb is extremely faint now and appears to be quite a bit farther out than it was before. . . . We're looking at the new little bright spot, and noticing that it has the appearance this morning of being a new active region. . . . We do see one small front spot associated with this region which I believe warrants being called an active region at this time. And we decided to call it Henry." Toward the end of the first mission, Kerwin wrote his observations in verse:

For the [scientists and engineers] and all those people,
It's a final debriefing from the Skylab crew:
Oh, our mugs are filled with [beer] of solar science,
While a Velcro strap* was holding down our rumps,
As with [instruments] in hand,
The monitor we scanned;
We were tracking down some dark and bushy clumps.
Yes, we've dreamed of active region seven-seven
'Til the panel's worn our fingers down to stumps.
And there ain't but one phenom
That you could always glom—
They are bushy, they are dark, and they are clumps.
Now, if suddenly the flare alarm goes crazy,
And it looks like Sol has got the mumps,
Forget X-ray; forget plage;
That stuff's naught but a mirage.
Fix your eye upon the dark and bushy clumps.
Then [back home] we'll sit around the table
To look at all the film and take our lumps.
And they'll say, "Why did you leave TONE/LIGHT enabled?†
We've got fifty thousand frames of bushy clumps.

Whatever the bushy clumps might signify, this was the
first poem written in space.

At twelve o'clock, the astronauts knocked off for lunch,
which was usually sandwiches. Gibson, who had been
sun watching in the docking adapter, slithered through
the hatch into the workshop and floated into the dining
room, for the astronauts always made a point of eating
at the same time; otherwise, they might never be together. They tried to keep to regular hours for their
meals; if they didn't, they found that their lives became
chaotic. The second crew in particular had run into

* Kerwin was still using the solar console's chair, which had a
Velcro seat belt.
† "TONE/LIGHT enabled" is a setting in which all the solar instruments are left running in order to record the maximum information, as in photographing a flare.

trouble. "We found the first three or four days that we tended to let the meals move a little bit," Bean had confided once to the B channel. "It became obvious after [a while] that there was enough work up there . . . to work all the time and not *ever* eat. I think that kind of upset us a little bit and probably was responsible in some degree for the fact that we were upset the first few days. So when we stopped letting the food move around, things kind of stabilized out for us." Even then, though, Garriott, who had a sort of boy scout's enthusiasm for anything scientific, had been apt to bring the conversation around to the exciting things going on on the sun, and he was often so persuasive that one or another of the second crew grabbed his sandwich and floated up to the docking adapter to spend the rest of his lunch hour on the console. This zealousness, of course, contributed to the accelerated pace Mission Control set for the third crew. Despite the example set by the second crew, though, the third crew preferred to keep their mealtimes for themselves. When the ground, in an effort to get the third crew caught up, began *scheduling* experiments during meals, these astronauts complained bitterly. "There's nothing worse than having to gobble your meal in order to get some task done that really should have been scheduled at some other time, so please, you know, I think in the future, if you would please, loosen up!" Carr said toward the middle of the third mission. "On the ground, I don't think we would be expected to work a sixteen-hour day for eighty-five days, and so I really don't see why we should even try to do it up here."

Mission Control had been turning out schedules for the men as if they were inputs for a computer. Indeed, Hutchinson, the lead flight director, prided himself very much on the way the ground had learned, during the first two Skylab missions, to keep the astronauts working. "Back at the first mission, we weren't good enough to schedule the guys tight, but by the time the second mission ended, we knew exactly how long everything took," he said in a sort of postmortem after all three missions were over. "We knew how long it took to screw in each screw up there. We could have planned a guy's day without leaving a spare minute if we wanted to—we had that ability. We prided ourselves here that, from the time the men got up to the time they went to bed, we had every minute programmed. The second crewmen made us think this way. You know, *we* really controlled their destiny."

There were those on the ground who saw what was happening and tried to stop it, and chief among them were the flight surgeons who, along with the capsule communicators, were the men in the control room most apt to argue on the astronauts' behalf against the flight controllers. The flight surgeons felt somewhat compromised in their efforts, though, because many of the extra experiments on the third mission were medical ones, so that they were themselves among the chief offenders. This, however, was not the flight surgeons' main obstacle. Dr. Jerry R. Hordinsky, the crew surgeon for the third mission, said later, "At conferences, when we were on the side of easing up, of saying that the flight plans were too much, the engineers couldn't understand what we meant. We witnessed Mission Control getting off on the wrong foot, but there was no place to blow the whistle. And communication between us and the flight planners was not good. They told us that 'the flight schedule was a nonmedical duty,' so there was a bad interface. It took us three weeks to see what was going on; then we went to bat." It wasn't until the sixth or seventh week, halfway through the mission, that the situation improved, and then it wasn't because of any initiative taken on the ground but because the astronauts themselves revolted, and Carr, the commander, told the flight controllers, in a memorable bitch, that he had had enough.

The third crew went on strike at the end of the sixth week. One day, Carr, Gibson, and Pogue stopped working and did exactly what they felt like doing. Gibson spent the day on the solar console, while Carr and Pogue sat in the wardroom looking out the window. They took a lot of pictures, not of scientific targets but of things they wanted to take. Carr made a sort of declaration of independence to the ground. "We'd all kind of hoped before the mission, and everybody had the message, that we did not plan to operate at the second crew's pace," he announced into the B channel. "And I think really up here my biggest concern is keeping the three of us alert and healthy. I think an illness is probably something that we can really do without. I'm also getting the feeling from some of the questions that have been asked of us the last few days that people there are beginning to hassle over who gets our time and how much of it. . . . We're beginning to get just little hints and indications that we're getting into a time bind—that it's got people really worried down there. People are asking about experiments, and the medics are asking about exercise, and [do we really need as much as we're taking], and why, and all that. And I'd like to know just exactly what everybody's motives are when they're asking these questions. . . ." Carr evidently realized that Mission Control itself was under pressure from NASA administrators who wanted results and from the scientists whose experiments were aboard, for he went on in a more conciliatory vein: "I imagine you guys are probably caught right smack in the middle of it, and the question that arises in my mind is, are we behind, and if so, how far? Or is all this hassle over our time a result of people coming out of the woodwork with new things to be done? . . . That's essentially the big question, you guys, and that is, where do we stand? What can we do if we're running behind and we need to get caught up? What can we do that's reasonable, and we'd like to be in on the conversation, and we'd like to have some straight words on just what the situation is right now. Commander out."

Carr's diatribe got results, for it persuaded the ground to give the astronauts fewer experiments to do and more time to do them in. The air was cleared somewhat, and the third crew's performance soon improved to the level of the second's, even surpassing it in such areas as sun watching. Carr blamed himself for not having blown the whistle on Mission Control sooner; after all, the director of the Skylab program, Schneider, had told the third crew not to work at the second crew's pace, an instruction that Carr had had every intention of carrying out. He said later, "We had told the people on the ground before we left that we were not going to allow ourselves to be rushed; yet we got up here and we let ourselves just get driven right into the ground! We hollered a lot about being rushed too much, but we did not, ourselves, slow down and say to hell with everything else and do things just one after the other like we said we were going to do." Now they had. A free exchange of ideas between the astronauts and the ground, of course, had been strained from the start by the episode of Pogue's nausea and the astronauts' failure to report it. Dr. Musgrave, the astronaut who as one of the capsule communicators had felt that this initial error was at the root of a great deal of the third crew's subsequent troubles, believed the episode made Carr, Gibson, and Pogue feel they had to try harder to make amends; it put them at a disadvantage when it came to complaining about their work load, and it may have made the ground a little less sympathetic.

After lunch, Gibson left the other two astronauts sitting at the wardroom table and went on up to the docking adapter to put in some time on the earth resources experiments package, a battery of five instruments underneath Skylab that would photograph the earth in different wavelengths. First, though, he had to put fresh film or tape in the cameras, which were set through the hull. Threading the tape was hard because it tended to uncoil more easily in space, and the end undulated back and forth like a snake. The astronauts were afraid to slam the door at the back of the instrument for fear of breaking the wriggling tape, and they grumbled that in the future, film and tape should be supplied in cassettes that could be slipped into place, instead of on reels that had to be threaded. Next, Gibson changed the space station's attitude, by means of the huge spinning gyros, so that Skylab no longer flew with the top of the telescope tower constantly toward the sun, but instead flew with its underbelly, where the earth resources instruments were, always pointing at the ground, so that each time the space station orbited once, it also rotated once. As the solar panels that generated electricity now rarely faced the sun either, the earth resources instruments had to rely on electricity from storage batteries, and therefore they could not be used for very long at a time. They had been added to Skylab as a sort of afterthought so late in the planning that they weren't integrated into the overall design as well as the solar telescope, which could be used constantly because it looked in the same direction as the solar panels. When NASA had first begun to plan Skylab, in the middle 1960s, before the Apollo trips to the moon paradoxically had helped kindle man's interest in his home planet, the agency had had very little interest in the earth, except as a launching pad. The idea of using manned spacecraft to study the earth had seemed a contradiction in terms. Skylab's gaze was to be fixed primarily on the sun, until the new interest in the earth's ecology and resources had forced NASA to accommodate it.

With Skylab correctly oriented, Gibson floated over to the earth resources console and anchored his feet in the triangular grid platform there. The whole docking adapter looked disconcertingly different, for he was now standing upright along the space station's long axis—a man standing at the solar console would be above his head and sideways to him, a situation the astronauts found annoying. His feet were now toward the hatch leading back to the workshop; between his shoes, Gibson could see the floor of the lower deck some 70 feet away. He felt as though he might fall. "If I step off this little platform, why, I feel I'm going to go whistling all the way down," he said. Gibson was one of the few astronauts to say he felt like falling in weightlessness—something that may have had to do with his own strong vertical sense. Before he could topple, Gibson pressed his eye into the eyepiece for targeting the instruments. Skylab was flying over vast sections of the earth never covered by astronauts before; as its orbit reached from fifty degrees north latitude to fifty degrees south latitude, it took in about seventy percent of the earth's surface. Yet Gibson didn't have much of a view of it from the docking adapter, where there were only a couple of narrow slits for windows; otherwise, peering through the eyepiece for the instruments, Gibson might as well have been in a submarine looking through a periscope. Indeed, the craft that Skylab most resembled was the *Nautilus*, the submarine in *Twenty Thousand Leagues Under the Sea* by Jules Verne—another ship that circled the world in an alien environment.

The earth was covered with clouds; during the first mission, Kerwin had said that the entire world was "socked in." Gibson was forever looking for rifts so that he could bore in on the ground. The astronauts had been given a long list of targets by the earth resources scientists, most of whom were now at these sites in order to get what they called "ground truth" to compare with the astronauts' data. The study of the earth from space was so new that the scientists were still learning the capacities of their instruments, which were far more sensitive than the ones flown in unmanned satellites, with film that had much better resolution. Consequently, the scientists did not always know the sorts of things the instruments, with their five wavelengths, might discover about the earth. Much later, for example, scientists studying the imagery from the radar altimeter, one of the instruments, discovered that readings of radar bouncing back from the ocean's surface coincided in certain respects with known major features on the ocean bottom, such as seamounts and deep trenches, and that therefore spacecraft equipped with altimeters could be used in the future for broad-scale mapping of the ocean bottom. From 269 miles up, huge features appeared that had never been detected before, such as faults and gentle domes hundreds of thousands of miles in area, which the earthbound had always been too close to see. Whenever these features proved financially rewarding, NASA, of course, was delighted. Geologists examining film taken over the southwestern United States discovered a fault that had shifted part of an oil field several miles away; once the slippage was found, the missing oil was, too. They discovered a new copper deposit in Nevada—copper deep in the ground emitted mercury vapor that discolored the surface, and this discoloration in turn was sensed by one of the instruments. Afterward, Carr and Pogue, who were zealous advocates for manned space flights, went around claiming that this discovery alone would pay for most of the Skylab program.

NASA, though, put the lid on such loose talk, for the respective merits of manned and unmanned space flights were still unresolved. Although NASA's preference was clearly the former, the agency knew it might have to settle for a larger share of the latter, and it didn't want its astronauts mucking things up. Initially, the earth scientists, like the solar physicists, had been skeptical of the value of men in space, and some things that happened aboard Skylab fortified their view. All too often the astronauts missed targets, or were doing something else when they were supposed to be photographing; and the scientists, who were out around the world getting their ground truths, sometimes had harsh words for the data they were not getting. At these moments, they, as the solar scientists sometimes did, wished Skylab was an unmanned satellite, for machines, not being human, had great regularity and reliability, which was particularly valuable when it came to frequent, repetitive coverage of a site. The astronauts, of course, felt that no unmanned craft could make up for their ability to bring back the film from the cameras or to repair the instruments, all five of which broke down at one time or another. In response, the scientists recommended unmanned satellites that could be serviced by astronauts in the space shuttle who, after repairing the satellite and recharging its batteries, would also bring back it's high-resolution film. The astronauts, who didn't like being relegated to the role of maintenance men, retorted that no machine could replace their abilities to take advantage of targets that came up suddenly, to shoot around clouds, or to comment on what they were seeing. The ground, they thought, was forever trying to downgrade them as human beings. Whatever the case, the scientists afterward were delighted with the pictures the astronauts had managed to take, together with their impressions of the earth from space. Six months after the last mission was over, scientists were still sifting through the upward of sixty thousand pictures, looking for data about ice movements, the migrations of fish, volcanic eruptions, melting snows, air pollution, water pollution, floods, droughts, sand dunes, sea conditions, weather conditions, hot areas in the ground that might provide geothermal power, and floating seaweeds that might be a source of food—a cross section of a variety of conditions on the entire planet.

The best place to observe the earth from inside Skylab was not the docking adapter with its tiny slits of windows, but the wardroom. When the astronauts were there, the earth appeared dynamic and alive; it was the view from the big wardroom window that convinced all nine of the Skylab astronauts that the earth had to be observed directly, as any living object should be, with all the flexibility and intelligence that a man could provide, rather than indirectly, by an unmanned satellite, as though it were dead and static. Part of the earth was always framed now in the round window, as though the astronauts were looking through the aperture of a microscope at a living tissue—all greens, blues, yellows, and browns. "I gained a whole new feeling for the world," Gibson told a visitor after he came back. "It's God's creation put before us, and whether you are looking at a bit of it through a microscope, or most of it from space, you still have to see it to appreciate it." Like the sun, the earth was an ever-changing kaleidoscope. Pogue said when he was back in Houston, "Every pass was different. It was never the same from orbit to orbit. The clouds were always different, the light was different. The earth was dynamic; snow would fall, rain would fall—you could never depend on freezing any image on your mind." The astronauts could take all this in visually better than the earth resources instruments with their narrow fields; looking out the round wardroom window, an astronaut felt like a scientist who had suddenly substituted a lower-powered, wider-angled lens on his microscope, the better to view the movements of an entire organism.

The most direct view of all, though, was from *outside* the space station, where an astronaut felt there was nothing between him and what he was looking at—as though he had slipped down the barrel of the microscope and was walking about the slide, magnifying glass in hand. "Boy, if this isn't the great outdoors!" Gibson said the first time he went out. "Inside, you're just looking out through a window. Here, you're right in it." And Lousma had said, after his return, "It's like a whole brand-new world out there! Your perspective changes. When you're inside looking out the window, the earth's impressive, but it's like being inside a train; you can't get your head around the flat pane of glass. But if you stand outdoors, on the workshop, it's like being on the front end of a locomotive as it's going down the track! But there's no noise, no vibration; everything's silent and motionless; there are no vibrations going through your feet, no wires moving, nothing flapping." Skylab was moving down the track so fast that Lousma had actually *seen* the earth roll slowly beneath him; it was so big, though, that he could barely make out its curvature, unless he was looking at the horizon. "It's a big ball," he had said, during the second mission when he had gone outside with Bean. "You know what's neat? As I look here, I can see the horizon move: a little blue line is dropping as we drop behind the earth. It makes it seem like a planet instead of just a picture, like you're really going around something. God! I'm glad that Saturn rocket worked! It's great to be here." In spite of his collegiate ways, Lousma was the most easily moved of all the astronauts, and the least afraid to say so.

The first astronauts to go outside were Kerwin and Conrad, who had gone out to deploy the solar panel that had been pinned to the side of the workshop when the insulation had been ripped off the space station by the wind shortly after launch. Sunlight blazed into the open hatchway—in the collar just forward of the workshop, where it met the docking adapter—with a brilliance they were unaccustomed to inside the ship. Outside, they could see the white scaffolding of the telescope tower looming above the docking adapter like a windmill; its array of solar panels, spread overhead like vanes, quartered the workshop with its shadows. Later, Kerwin would be going up to the top of the tower, which the astronauts called the sun end, in order to change the film in the solar cameras there. All the astronauts who went up there agreed that the sun end was the most exhilarating place aboard Skylab. "To be on the end of the telescope mount,

Lousma under the telescope mount: "Its like being on the front end of a locomotive as it's going down the track! But there's no noise, no vibration; everything's silent and motionless" *(facing page, top)*.

The view near the base of the telescope mount: Skylab was moving so fast that Lousma could actually see the earth roll by beneath him *(facing page, bottom)*.

Lousma on the sun end: "To be on the end of the telescope mount, hanging by your feet as you plunge into darkness, when you can't see your hands in front of your face—you see nothing but flashing thunderstorms and stars—that's one of the minutes I'd like to recapture and remember forever" *(above)*.

hanging by your feet as you plunge into darkness, when you can't see your hands in front of your face—you see nothing but flashing thunderstorms and stars—that's one of the minutes I'd like to recapture and remember forever," Lousma had said afterward. It was a little unnerving, too. When Pogue went up, he had the uneasy feeling that comes with being in the crow's nest of a ship. The telescope tower didn't sway like a ship's mast; it was just that an astronaut up there was far enough away from any large structure that he no longer felt part of the space station. An astronaut anchored himself there by stepping into a pair of golden shoes, and if Gibson, standing in the shoes, leaned over backward, he felt he might fall off. "Hanging from your heels, it's just between you and the earth below, and you have the feeling that there is gravity, and down you could go," he said. Gibson, who also worried about plummeting feet first through the workshop dome, suffered more from acrophobia than any of the others.

Conrad had put together the sections of a long pole with a cutter on the end like a pruning hook; he would be needing it to sever a strip of aluminum—the remains of the space station's insulation—that had pinned down the solar panel. The strip prevented the panel from springing perpendicularly out from the workshop's side, where it belonged. Then, smoothing out the rope that would allow him to operate the blades from a safe distance, he started out the door and began moving slowly toward the undeployed panel across the huge cylinder of the workshop. In weightlessness, he had no traction on the rounded surface, and he continually felt as though he would float away. He couldn't float far, of course, because he was tethered by his long umbilical, his lifeline leading back into the space station. Behind him, Kerwin untangled Conrad's umbilical from his own as he steadied the long pole, one end of which was attached to the telescope tower behind him, so that Conrad could use it as a handrail.

Kerwin called Conrad's attention to the sunset, an arc of rainbows over the horizon; it looked especially pretty through the big solar panels spreading from the top of the telescope mount overhead. The astronauts would have thirty minutes of darkness when they were unable to do any work. They were impressed with how brightly lit up the earth was at nighttime. Looking down through the moonlit clouds, Conrad could see hundreds of grass fires surrounding what he took to be villages in Africa, where there were more grass fires than anywhere else on earth. On all sides, lightning illuminated patches of thunderheads. He thought of his return from the moon when, from a distance of twenty-five thousand miles, he had seen lightning flashes ripping up the entire nighttime side of the earth. What most astronauts remembered best about these periods of darkness, though, was simply *being* there, with nothing, absolutely nothing, between them, the earth, and the stars.

When it was light again, so that Conrad, out by the broken panel now, could get on with his repair job, he attached the cutter blades to the strip of aluminum that was holding it down. He also attached a rope directly to the panel itself, so that if it didn't spring properly into place he could pull it upright without having to get near it. At a warning from Weitz, who was watching from inside the docking adapter, he flicked his umbilical away from the hinge that attached the panel to the workshop, where it might have gotten pinched. Then he retreated halfway back to the telescope tower, for the solar panel, when the aluminum strip was cut, could well spring up with some force. He steadied himself with the long pole. At the other end, back by the telescope mount, Kerwin began hauling on the cutter's own rope, exactly as if he were using a pruning hook. After several tugs, he man-

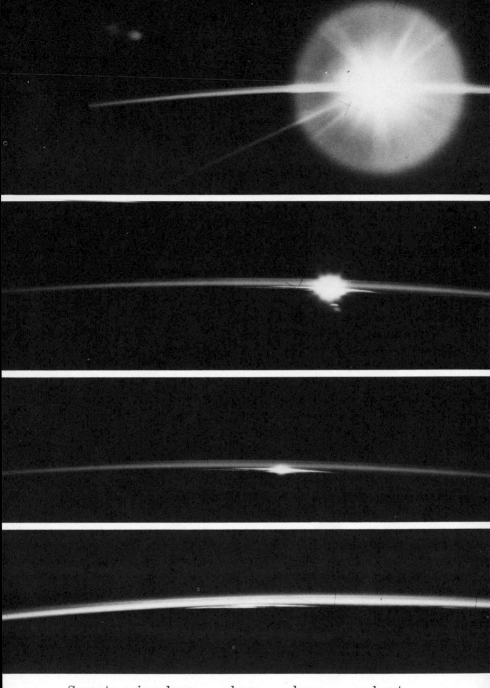

Sunset: going down . . . down . . . down . . . and out.
Sunsets, and also sunrises, were accompanied by a vivid
arc of colors like a rainbow stretching across the horizon.

aged to sever the aluminum strip. Now that the long pole wasn't being held down anymore by the cutter, though, it floated into the air, and Conrad floated along with it; he had to pull himself hand over hand toward Kerwin, who still had hold of the other end. When they looked back at the panel, they saw that it had only popped up a couple of feet; it still wasn't locked at right angles to the space station. The two astronauts hauled on the rope attached directly to the panel, but they couldn't raise the panel any higher. Kerwin said he would stay where he was, braced against the airlock door and keeping tension on the rope, and he suggested that Conrad go back out to the beam and shove up underneath it with his shoulder. Conrad reported later, "I gave a mighty heave, and the science pilot gave a mighty heave, whereupon everything went black and I shot straight up into the air. And as I revolved to my right, I could see the beam coming out at a good, smart clip. And by the time I settled down from my whifferdill back in the airlock area, the beam had come all the way up and was fully deployed." Conrad said to Kerwin, "Would you look at that!" and Kerwin replied, "I expected it to come, but I expected to lose you, too. By gosh, we got you and the solar panel."

Usually, though, the astronauts did their earth watching from the relative calm and security of the wardroom, where they had a better chance to study their planet. Though other astronauts had seen the earth from space, none had done so as extensively or in as great detail. Sometimes the Skylab astronauts had to clear bits of their dinner off the window before they could look out. They couldn't clean off a spot of ice that had formed outside in the center of the glass, though, and this annoyed them. If the view from outside had been intoxicating, the view from the wardroom was merely addicting; indeed, Lousma had been reminded of the compulsive way he had first watched television as a boy—by pressing his nose against a store window. The wardroom window soon became smudged with nose and fingerprints, and as the glass was cold, the astronauts insulated their foreheads with bits of cloth. They preferred to look at the earth with the horizon horizontally in front of them, as though they were standing on the ground. Under certain circumstances, though, the horizon slipped around the window so that they had to move to keep it in view. Gibson noted one day, "You might start out with your feet on the floor, and a little while later you'd look back and find the other two guys were upside down, sitting at the wardroom table, which was upside down, too." (When this had happened to Kerwin, he had felt he was in Wonderland.) An earth-watching astronaut had to be something of a contortionist, jamming himself into odd and painful positions between the window and the ceiling or the window and the walls. The third crew recommended that in the future spacecraft windows should have a margin 6 feet wide around them, so that an astronaut could perch comfortably anywhere around its circumference, and that there should be more grips to hold themselves in place. They wished the window was bigger, and that there were more of them. As windows structurally are weak points in a spacecraft's hull, designers are not apt to follow this advice. Even so, Gibson recommended that aboard the space station he was continually planning there be a transparent observation bubble slung beneath, just like the turret a gunner sat inside aboard a bomber. There, an astronaut would have the world laid out around him, and he could quickly swivel some high-powered instrument at any object that caught his attention. The observation bubble, he felt, should be manned twenty-four hours a day.

Still, the astronauts in the wardroom had a better view of the earth than any others who had orbited before. Not only was the window, at 1½ feet in diameter, bigger than in previous spacecraft, but the space station was flying over a greater portion of the earth, crisscrossing the world in vast sweeps that might take it from southern Canada to southern Africa through China and Russia across the Pacific and over the tip of South America. Sometimes the astronauts simply gawked, like any sightseers. Occasionally all three members of a crew would look together, pointing out cities and peninsulas as uncertainly as though they were schoolchildren learning geography. "Yeah, we're coming right down into Greece," Carr said from the window, "and that island, out there, I believe is part of the Peloponnesian—what's the name of it—the Dardanelles? Big World War I naval battle down there." There was no place like home, though, and Carr and Pogue were particularly attentive whenever they were anywhere near Texas or Florida, where there were a number of NASA and Air Force bases where they had been stationed. On these occasions, they talked almost as fast as the world rolled by:

"OK, come quick—you can see the whole Texas coast from Brownsville to Houston, Beaumont, Port Arthur—oh!" Pogue called out.

"Oh, heck. Look at that," Carr said, joining him.

"There's Brownsville, and around the coast you can see Galveston Bay outlined by the darkness," Pogue said.

"Yes," Carr replied.

"San Antonio, Austin. Let's see if we can see Fort Worth."

"OK."

"I can't quite crane my neck far enough. . . . Trying to see the Astrodome. . . . See New Orleans down there? Gosh, what a tremendous view. The whole Gulf Coast looks clear as a bell."

"Yes. . . . See the [oil] wells on the water."

"Think you're going to pass just about over Pensacola," the CapCom cut in.

"Yes," said Pogue, "we have the whole Florida penin-

CARR: "Yeah, we're coming right down into Greece, and that island, out there, I believe is part of the Peloponnesian—what's the name of it—the Dardanelles? Big World War I naval battle down there." Sometimes the astronauts had a hard time figuring out where they were. (Greece, *above*; the Dardanelles, *right*.)

sula in sight. All the way down to the Keys. Miami's lit up. Look at Miami Beach! Tampa! St. Pete's. . . ."

"Oh, boy," Carr said.

"You can see the whole thing!" Pogue said.

"Just down there around the Everglades, it's the only place that there's no light."

"Tallahassee! That's right."

"Tallahassee . . . Atlanta! Holy cow! Look at that! You can see the interstate highway all the way down the center of Florida."

"That's right."

"Right down the middle of it like a backbone."

"There's the Cape, Cape Kennedy. See the Cape plain."

"Yes."

"Orlando."

"Merritt Island—the whole smear!"

"Cocoa Beach!"

All the continents looked very much alike, for they had the same mix of plains, forests, mountains, and deserts. About the only place they could recognize at a glance was the southern end of South America, which tapered unmistakably. Sometimes Carr found that he could tell where he was by studying sand dunes, huge ripples some twenty miles long that were characteristic shapes in different places; in one desert, the dunes were star-shaped. Pogue liked the Sahara best, for its reds, golds, oranges, and browns were most brilliant. Lousma had thought the whole world seemed made of sand. "Yeah, here we're coming up on the Gold Coast of Africa," he had said during the second mission. "Mighty barren country down there; looks like lots of desert sand dunes, very pretty colors, all various shades of brown. I'm just going to have to take time out and take a picture of this, folks. I mean, we probably got four hundred pictures of Africa, but everyone seems impressed every time we come up on this area. . . . Here we're passing over Saudi Arabia. We get around. Lots of sand down there, wow!" If there was one thing there was more of on the earth than sand it was water. When the astronauts were over oceans, which contained no characteristic shapes or identifying features, it was almost impossible to tell where they were, and they were over oceans about seventy percent of the time. "You don't realize how much ocean there is on earth until you see how much time you're looking at water," Carr said after he got back. During the first mission, Weitz, that crew's pilot, had not been able to get over what a vast expanse the Pacific was. "You don't comprehend it by looking at a globe," he said. "But when you're traveling at four miles a second, and it still takes you twenty-five minutes to cross it, you know it's big." He had spent the time looking for atolls.-

The astronauts were continually surprised at how much time they spent looking not only at oceans and deserts but also at snowfields and mountainous areas, in none of which could they see any sign of life. In contrast to the Apollo astronauts who had looked back from the moon and described it as an oasis in space, the Skylab astronauts thought the earth a barren place. The toughest part of the earth to survive in that they passed over, the third crewmen thought, was the area from Tibet across Outer Mongolia. "There is nothing but a great big nothing out here now," Gibson said during the third mission. "Northern China, Outer Mongolia, and all that gold stuff: the Gobi Desert." Carr, especially, thought man had a tenuous foothold on his own planet, where the checkerboards of his cultivation seemed to be packed into the few temperate areas, or fringed the deserts and oceans like a green mold struggling for existence. "Not much of the earth is hospitable to man," he radioed down one day—as though Mission Control's presence there had somehow made it seem an alien place. "We don't occupy much of our world. We're crowded into small areas." Carr seemed to think more about man's relationship to his planet than did any of the others. Like his cultivation, man's cities seemed to splotch the earth in giant patches that stretched long tentacles as if they were cancers drawing nourishment from a meager host. Life on the oasis struck Carr as hard—perhaps the way life in space was for his crew. From his perch in space, he didn't think that his species was thriving; man seemed to him to have made very little mark on his planet. Carr saw no signs of what might be called progress; he had about as good a chance of seeing an Asian dirt road, he said, as an American interstate highway because the former was very apt to be chalked in thick lines by the clouds of dust rising from it.

Whatever the astronauts thought of the earth during the day, it was a different place at night. Every ninety-three minutes, as they plunged around to the dark side, they could see the lights of cities twinkling like jewels; the lighted streets of Mexico City were laid out in such a way that Carr, who didn't like the look of cities in the daytime, thought it seemed like a five-pointed star. Others were fooled by the twinkling cities. During the second mission, Garriott had exclaimed, in some confusion, "There's some lights! They could be stars. I don't know. No, they're cities. There's one little town going by. It looks like a little bug from here. It's just flying by. And there was a flash! I don't know whether it was one of those flashes like you see in the eye up here, or whether it was some sort of meteoroid."

Frequently, the second crew discovered, the nocturnal fireworks were in the sky. On one occasion, Lousma had called to Garriott, "Hey, there's an aurora, Owen; come here! Hey, Owen, there's a big aurora here."

"Say, amazing. I'll be right there," Garriott had called back, without coming.

"I don't think that's the sun coming up, 'cause it's the wrong direction," Lousma said as Bean came up. "Look at it, right there."

"Yeah, that's an aurora," Bean said.

"Isn't that a beauty?" Lousma said.

"Oh, it's pretty," Bean replied. "You get any pictures of it?"

"Take a look at that aurora out there! Right out here," Lousma said to Garriott, who came by at last.

"My God, look at that aurora!" Garriott said.

"Isn't that pretty?" said Lousma, who didn't seem able to leave his aurora alone.

"That's a beauty, Jack," said Garriott. "God, that is really good."

The next morning, Garriott, never the one to waste an opportunity for a scientific discourse, dictated, "I want to comment on this aurora that we saw last night. The majority of it was closer to us than the earth's horizon. It was greeny in color; I couldn't see any red. The arc extended out away from the spacecraft toward the horizon, and the aurora tended to blend in with the airglow. Now above it, very faintly, you could see streamers, very thin striations in the aurora extending from much higher altitudes. Very thin rays, very dim but thin rays more or less vertically aligned with the magnetic field could be seen. . . . It was probably the most extensive auroral display we've seen."

The night always passed quickly, blazing suddenly into daylight with a sunrise that looked like the beginning of an old Warner Brothers movie. Lousma had said once, "A guy like me, who likes both sunsets and sunrises, mostly gets to see sunsets. But here, in space, every day we get sixteen of each." Whatever they were doing, the astronauts frequently crowded around the window at such times. Once Bean had informed Garriott that the sunrise was beautiful, but Garriott, who had not found the word adequate to the aurora, urged Bean on to greater flights of description; and Bean, who was the only one of the astronauts who didn't enjoy looking out the window, and who was by no means God's gift to the English language, did his best. "There's a dark blue streak and then that sort of gold, orangy-gold color," he said. "When the sun first begins to come up, it starts a little bit blue with the littlest bit of gold and then it just gets bluer and bluer, and the gold gets steadily a little bigger, and then all of a sudden the gold starts getting real wide and then the sun comes out."

Garriott had been so enthusiastic about the view from the window that the third crew was, as usual, given a heavier load of things to look for and questions to answer. ("Garriott was forever looking out the window!" Bean, the only one of all the astronauts who wasn't doing so, had said during the second mission. "Now, I couldn't any more look out the window all that time for two months for anything!" Bean, ever practical, had put it down as one of the quirks of having a scientist aboard.) The scientists on the ground were interested in the sorts of things the men could tell about the earth from space, just as they had been with the earth resources instruments. The Skylab astronauts were the first ones to look at the earth systematically and to try to describe what they were seeing, and the third crew gave the fullest descriptions; Carr had the sharpest eyes and gave the most detailed observations. Gibson wished he had some high-powered equipment in the wardroom. One of his biggest complaints about Skylab was that the earth resources instruments, which were very powerful, were in the docking adapter, where it was hard to see out, while down in the wardroom, where the best view was, there were only two hand-held cameras and a pair of binoculars whose focusing knob didn't work well. "Sometimes we saw a fault zone that we hadn't known was there," Gibson said on one occasion. "It would have been nice if we could have quickly gotten some instrument on it. Or ice islands, for example, which came up from Antarctica and which appeared unexpectedly. Or the wakes that mountainous islands made in the clouds. Or a sudden glimpse of pollution in Mobile Bay. You could take advantage of all these things if a man were up there, properly set up. Aboard Skylab, an observer of the sun at the telescope console could do this, and there's no reason why we shouldn't look at our own planet in the same way. We have looked at the sun for years, and the moon, but we have never really looked at the earth before, and that shouldn't be." Presumably Gibson's underslung plastic dome for his future space station, with its high-powered optical equipment, would remedy the situation.

As time went on, though, Gibson, Carr, and Pogue developed all sorts of tricks to help them see better: they found they could see coastlines best if they waited until the sun's reflection glinted off the water by the shore. They learned to find ocean currents by differences in the clouds above them. They learned to look below clouds, if there were breaks in them, and see most of the ground underneath, as the space station flew along, as though they were looking through a picket fence from a moving car. The scientists asked them to look for faults and domes in the ground; for volcanic eruptions and how the wind blew the smoke away; ocean currents that were apt to be different colors from the surrounding water; and upwellings of plankton from the floor of the ocean, which could be spotted on the surface as green smudges called plankton blooms. There was always something going on:

CARR: "Spacecraft was located up over the North Pacific. We were looking to the northeast, where we could see the Canadian Rockies. We could see great masses of stratus clouds stacked up behind the mountains, and the mountains were sticking up through the clouds and causing great [disturbances] in the clouds along a very wide front from what looked like Alaska all the way down to almost Washington State."

GIBSON: "We were able to see the plankton blooms resulting from the upwelling off the coast of Chile. The color was light and contained a fair amount of green in contrast with the dark blue [of the ocean]. The bloom itself extended along the coastline and had some long tenuous arms reaching out to sea. . . . The arms or lines of plankton which were pushed around in a random direction, fairly well defined but fairly weak in color, contrast with the dark blue ocean. . . . We could not observe any darkened water which might represent the upwelling itself, relative to the ocean water. So the dark blue which we saw we could not contrast to the ocean water farther out. It was kind of an impressive sight, though. The fishing ought to be good down there." [Music.]

CARR: "I took a picture of an ice island. It was several hundred miles east of the southern tip of South America. What it looks like is a great big slab of ice that broke away from the pack ice. And there's no mountains or crevices or anything in the ice. It just looks like a big sheet of ice that's floating free. And floating in the water around it are many, many small pieces, small chunks of ice. And I don't know if those are what you call icebergs or not. I don't know how much of them are sticking up. But they are rather small."

POGUE: "The pilot again reporting on the discoloration of the ocean: I thought I saw some brownish spots in certain areas. And Jerry [Carr] was looking on farther south of the Falkland toward the Strait of Magellan, and the Cape of Good Hope, and we started to see some reds or some telltale marks of petroleum pollution so it could be that."

CARR: "I was looking out the window down at the earth at night, and I saw a great huge patch of fires. Looks like they're located in the area of Nigeria, and Chad, and Cameroun. It was a long, elliptical area of many, many fires. I would guess that it's probably two or three hundred miles long, and probably fifty to sixty miles wide, and there were many, many, many fires. I wish that I'd had a camera with the right kind of film in it. I think it would have been an excellent picture. Off to the south of these fires, there was an overcast with many thunderstorms, and many, many bright flashes of lightning going on at the same time. It was a spectacular night scene: I presume that the fires we're looking at are [lightning] flash burnings."

GIBSON (*at night*) : "We've just been coming over, in the past five minutes, an extensive area of thunderstorms. . . . There seems to be some sort of a collective organization to the lightning strikes which occur over a wide area. When one goes off, two or three may go off simultaneously, or one of those may turn out to trigger a whole lot of other ones all over a very, very wide area—five hundred thousand square miles, perhaps. The lightning flashes then will go off, numerous ones, ten, twenty, forty, fifty. It'd be calm again for about one to two seconds, then we'd get another period of—oh, maybe three, four, five, maybe up to seven seconds or so of lightning going off in all locations. And it subsides; period of calm; then cycle through that again. A few things which impressed me here: one is the fact that they could go off simultaneously or near simultaneously over a large distance—sympathetic lightning bolts, if you will, as analogous to sympathetic flares on the sun. And that we do get periods of calm between periods of very high activity. Some sort of collective phenomenon appears to be at work."

CARR: "At 01:55, I happened by the wardroom window just in time to see us passing over Korea, with the

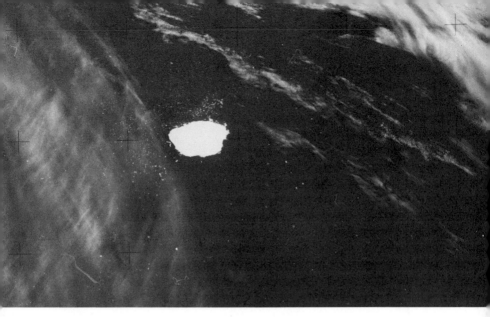

CARR: "I took a picture of an ice island.... It just looks like a big sheet of ice that's floating free. And floating in the water around it are many, many small pieces, small chunks of ice."

CARR: "I decided to take a look to see if we might get a good look at Sakurajima, the erupting volcano in southern Japan.... There was a very heavy, very large pall of smoke blowing out of the volcano and then blowing horizontally off to the southeast."

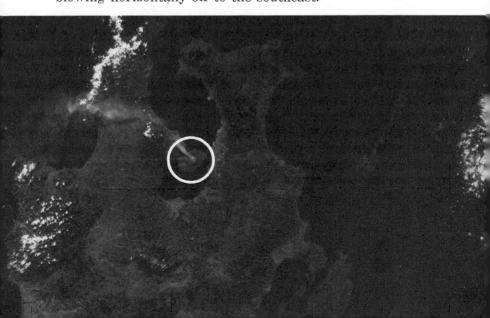

weather quite clear, and I decided to take a look to see if we might get a good look at Sakurajima, the erupting volcano in southern Japan on the island of Kyushu. And it was indeed clear there. There was a very heavy, very large pall of smoke blowing out of the volcano and then blowing horizontally off to the southeast. . . ."

CARR: "Now as we were crossing the eastern coast of South America, I looked out to sea, and I got an absolutely perfect [view] of the confluence of the Falkland and the South Equatorial currents. You can see the Falkland Current coming up from the south, and you can see, not quite so clearly, the Equatorial Current coming down from the north. And right straight out from the city of Buenos Aires, you can see where these two currents meet and head straight out to the southeast. The colors down south [of the Falkland Current] are more of a powdery aqua, toward the chartreuse; whereas the [Equatorial Current] coming down from the north are more of a deeper blue, turquoise or aqua. Where the two meet and head out, you have a mixture of the two colors, and it is very streaky, very taffylike, and serpentine. The water itself is a good blue."

CARR: "Photographing that lake?"
POGUE: "Yeah, I want to see if I might be able to do Lake Tahoe."
CARR: "Did you notice how Lake Tahoe looks kind of like a footprint?"
POGUE: "Sure does. Looks like a bear paw. . . . Oh, yeah, there's the Grand Canyon, beautiful. . . . There's Lake Powell."
CARR: "The old Grand Canyon, huh?"
POGUE: "Yeah. It looks huge."

GIBSON: "I'd like to just try to describe this tropical storm to you briefly. First of all, it was just, in general,

POGUE: "There's the Grand Canyon, beautiful. . . . There's Lake Powell."
CARR: "The old Grand Canyon, huh?"
POGUE: "Yeah. It looks huge."

A tropical storm similar to the one Gibson saw

a relatively small but compact, well-framed storm. . . . The prestorm squall lines were light and not very extensive activity along them. However, they certainly did define the vortex slope. Cirrus streets were present; they were also light and they also define the spiral slope. The eye appeared well defined. The sides were sloping gradually in and the eye was very circular. However, we had an oblique view to it and I could not see all the way down to the water."

"Well, enough of this lollygagging," Carr said. "We got to get to work."

"Yeah, darn it," said Pogue.

"What do you mean, Carr?" Gibson asked. "Don't we have two more hours of lollygagging?"

"We blew our lollygagging time yesterday," Carr said. "Now we got to go to work."

Dinner, which the astronauts normally had at about six o'clock, was the big meal of the day; it included meat, vegetables, and a dessert. The meat was usually veal or dried beef. The beef was the food the astronauts liked least; even the usually polite Garriott had said it tasted like insecticide. Nonetheless, they ate most of it. They were hungry; astronauts aboard Skylab tended to get hungry more quickly than they did on the ground—a fact that had mystified Kerwin, who wondered how it could be that in space, where the astronauts lost weight, and where they exerted themselves less than on earth, they still managed to consume the same amount of calories. As Apollo and Gemini astronauts had eaten less, the Skylab astronauts may have underestimated the calories they burned up getting about inside the space station, which was thirty-three times larger than an Apollo spacecraft and a hundred times larger than a Gemini. The third crewmen were particularly apt to be hungry, and stay that way, if the day was one of the days they had to eat a variety of food bars. They had brought these with them in order to stretch their supplies, which were meant to last only fifty-six days, to eighty-four; the bars were flaky and greasy; though they contained as many calories as a regular dinner, the third crew never felt filled up afterward. Most of the astronauts would have liked a drink with dinner, and indeed there had once been a wine-tasting party at the Space Center for them to pick the one they liked best; however, the Women's Christian Temperance Union had gotten wind of the plan and NASA had aborted the project. NASA will go to any length for public relations. Conrad, though, had been indignant. "It was all part of the idea of habitability, to make the place more like home," he said. "If folks drink a martini at home, why not up here?"

After dinner, one of the astronauts recorded for the dietitians exactly what each of them had, or had not, eaten that day, as though he were reporting to a particularly watchful nanny. Kerwin had done this one night during the first mission: "Hello, tape recorder. . . . So far, the commander has eaten everything except the corn, which was a big failure. The bag failed at its seam. Got corn powder all over everything. The science pilot has not had his biscuit and jam or his peach pudding. He intends to eat at least the peaches and the rest as he gets it. However, the science pilot will probably not get to his apricots tonight nor his butter cookies. The pilot only had half of his bread for breakfast. And has yet to eat his vanilla wafers or his lemon pudding from lunch."

After dinner, which was when the astronauts looked over the schedule for the next day, Carr went up to the docking adapter to collect the latest instructions from the ground on the teletype machine. Every day Mission Control sent up about six feet of instructions of things for the astronauts to do, and if they had had to copy them all down longhand, the way the Apollo astronauts had done, they wouldn't have had time to do any of them;

Garriott, in the wardroom, catches up on some desk work after dinner; he is glancing through the hatch to the upper deck.

Bean, in the wardroom, wrestled with some teletype tape that wouldn't stop rolling up. The window behind him is uncovered, the way it would be for viewing the earth.

the astronauts were as concerned that the teletype might break down as they were that the trash airlock might. Carr ripped the latest instructions off the machine and took them back to the wardroom, which in the evening was usually awash with teleprinter paper. In weightlessness it tended to curl at the ends; Carr held it flat on the wardroom table with a variety of magnetic paperweights, and when he ran out of these, he broke out a deck of magnetic playing cards and used them. When he ran out of cards, he used tape, but the tape was gummy. Clearly desks on future space stations, where there would be even more bureaucracy, would need some thought. When Gibson had experimented with turning the screens on the workshop ceiling into an aerodynamic workbench, he found that the moving air held down paper even better than it did hardware, and possibly future space stations will have aerodynamic desks.

The time from eight to ten at night was supposed to be the astronauts' own to use however they wanted, but matters never worked out that way. Mission Control always had more experiments for them after dinner. Rice seeds germinating in a box on the wardroom wall had to be photographed; gypsy moth eggs had to be counted to see if any of them had hatched; elodea leaves had to be examined under a microscope to see if the cytoplasm in their cells circulated readily in space; and specially prepared wicks had to be dipped into water, to see if capillary action worked without gravity. It got so that the astronauts had almost no leisure time at all, in spite of the fact that in the wardroom there was a sort of games cupboard, called the off-duty equipment assembly, which was filled with taped music, balls, darts, playing cards, and books. (Before the books—paperbacks selected by the astronauts themselves—had been allowed aboard Skylab, NASA had had to test them for flammability; some engineers at the Space Center had set fire to a number of books and found that—contrary to what a casual reading of history might indicate—they were extremely fire-resistant. They turned out to be what one engineer called "great ablators," for one page had to be on fire before the next reached its kindling point; books even flake in a heat-dispelling manner, like the ceramic shields of spacecraft.) The astronauts had little chance to read a book and less to use any of the other off-duty equipment, with the exception of the taped music that they played constantly. Once, when Gibson was asked how he liked the games, he replied, "Off-duty activities? You gotta be kidding. There's no such thing up here. Our days off, the only thing that's different is that we get to take a shower."

Carr's diatribe to Mission Control during the sixth week of his flight, when the astronauts had gone on strike, had resulted in the third crew's being given more time to perform experiments, but not in any marked increase in their time off; loosening up the experiments had simply been a way of utilizing the astronauts more efficiently. The ground had never stopped thinking of them as robots to be manipulated for maximum productivity; it did not understand that free time had a special value of its own.

Though the first two crews didn't care much about their free time, the third crew, which was up the longest, resented the way the ground filled its evenings with experiments. Carr, who usually concentrated his vituperations on the schedule, said, "I think it's becoming apparent to me that [our evenings] are being impacted more and more as the days go on. Tonight we had to do the ultraviolet airglow horizon photography, and the problem here is that right as of today none of us has as yet had any time to sit down and read or write or just stare out the window unless we do it after bedtime when the ground stops talking to us. Now I would like to request that these periods be kept just as open as possible in order for us to get some relaxation. . . . I think you'll find that you'll get better work out of us, we'll be more rested and more efficient if we can do the following thing: that is, at regular times have some time for just plain quiet relaxation with nothing bugging us, no requirements on our time—just a period of time to be quiet."

When the third crewmen snatched a few minutes off, they didn't play with the games; they preferred to relax, to read a little, or to look out the window. Skylab, though, wasn't particularly set up for relaxing. "There aren't many places you would want to spend a whole lot of time reading, because the light is so bad it hurts your eyes," one of the astronauts said. "During the daytime, the window is about the brightest place. But the wardroom gets pretty crowded during all the multiple functions we've got to do there: looking out the window, eating, using the tabletop for checklists, and all that stuff." Carr occasionally managed to sneak off to the command module—the most private spot in the entire cluster—and turn off the air-to-ground radio there in order to get a few pages read; he got halfway through his second book. He, Gibson, and Pogue also got a little reading done while they were lying in the LBNP barrel and even while they were walking on the treadmill, propping the book on the seat of the bicycle.

Actually, everybody except the third crew had lost sight of what was to have been the chief purpose of Skylab, and that was to see whether men could really *live* in space for long periods of time; NASA had originally construed the word *live* to mean decently, as one might on earth, with regular shifts and time off for relaxation. Indeed, a few scientists and engineers at the Space Center regarded the astronauts' leisure time as an experiment in itself, the success of which would be essential to any long-range plans to have men spend long periods of time in space stations, at moon bases, or on three-year roundtrip excursions to Mars. Hutchinson, the lead flight director, didn't see it that way, though. "The initial purpose of Skylab may have been to explore simply how to *live* in space, but the cost of the program—two and a half billion dollars—caused us to change our minds," he said after it was all over. If there had been a change of plans, no one had informed the third crew, nor, evidently, Schneider, the director of the Skylab program. This could happen because the flight controllers, who were responsible directly to the director of the Space Center and thence to James C. Fletcher, the NASA administrator in Washington, had, because of the risks involved in space flight, not questioned authority over the operations of a mission and, given the fast pace of the earlier flights they were used to running, they were not what they would have called "programmed" for anything less than an all-out effort. "Our system was designed to squeeze every minute out of an astronaut's day," Hutchinson said. "Suddenly the system is asked to stop for a few hours, or a day, to give a man some time off. The system doesn't want to do it! Say, for example, that there's a day off coming. But say that there's a perfect pass over an earth resources target, and that the only chance to get that target is on that day. Now, are we going to say to the scientist whose experiment it is, 'No, goddamn it, we're going to lose that real estate for the Skylab program?' " It was, somehow, typical of NASA to send men off on a totally new sort of experience, and then overplan it to such an extent that they had no time to think about what they were experiencing—an important reason for sending human beings into space in the first place.

What Hutchinson, and NASA, hadn't reckoned with was that the third crew was no longer guided by the same stars that still governed Mission Control. Of course, it would be difficult for a man to spend several months floating weightless in a new environment, where local verticals shifted from room to room and where he was first one way up and then another, and then come out with the same perspectives as when he went in; at the very least, he would be receptive to new points of view. The first two crews, of course, had scarcely been affected, but they hadn't been up as long as the third crew, they hadn't been under such continuous pressure, and they may have felt closer to each other and to Mission Control. Whatever the case, the third crew was ready for some new ideas. And Pogue, who had had the most difficult time in the space station, was the one who had them. "I came to realize, during Skylab, that what we were doing was taking a human and making him function in a way he was not designed to," he said. "We were trying to function at a higher level of efficiency than we could. I then proceeded to make errors and berate myself. Finally I came to the realization that I'm a fallible human being, that I cannot operate at a hundred percent efficiency, that I am going to make mistakes. When I tried to operate like a machine, I was a gross failure. Now I'm trying to operate as a human being within the limitations I possess. . . . I think a person needs to more or less re-create himself, to pause and reflect occasionally. . . . I think that in order to act creatively, you have to have certain periods of time when you have to just stop and think and see yourself, and be aware of the situation, and sort of involve yourself in the totality of the experience at hand. We've got to appreciate a human being for what he is." Pogue might not have used quite these words but for the fact that in his few off moments he had been sneaking off to the command module with two books of popular psychology, *Self Renewal*, by James Gardner, and *Man the Manipulator*, by Everett Shostrom. Nonetheless, his outburst—and indeed the entire third mission—was a sort of affirmation of man over machine; in this case, the machine was Mission Control, men who were the integrated components of an advanced technology, who were trying to turn the *astronauts* into machines, as indeed they had always done. This time, though, the mission was too long, and the men, as men will, resisted being robots.

During the long string of space flights that had begun with the Mercury program, astronauts had been thought of as mechanical supermen, something the astronauts themselves had half believed. If the third crew had thought this way at the beginning of the mission, and Pogue indicated that it had, it clearly did not at the end —a necessary adjustment, perhaps, if space flights are ever to become more routine. After the third Skylab mission, it is hard to see how astronauts can ever be the same again. In the past, others had asserted that they had changed as a result of having been in space—one Apollo astronaut had left NASA to become a missionary, another to study extrasensory perception, a third had had a nervous breakdown—but none had ever transformed so dramatically while they were right under the eyes of Mission Control. A year later, Pogue himself would leave the astronaut corps to join an evangelical group.

If the third crewmen were influenced by anything in space, it was not so much by being weightless or by floating upside down as simply by what they saw out the window; they preferred looking at the view to anything else they did in their free time. "It was a shame to read with all that going on outside," Gibson said. Though he read a little when he was over water, he invariably put his book down when Skylab reached the shore, and looked at whatever continent lay below. (He frequently lost his place, for when he put his open book face down on the table, its hinges would force it slightly shut, flipping it into the air; if he put in a bookmark, the hinges would open, the pages fan out and the bookmark flutter away.) As Gibson and his two crewmates sat looking at the earth, they found that they were being drawn into a new frame of mind. Much of what they saw they already knew, but actually *seeing* it gave it a crystal clarity. Gibson, for example, knew that the world didn't have boundary lines between countries marked on it like a library globe, but he was nonetheless surprised when he saw from space that there indeed were no dividing lines between peoples. "In no way could we on earth, or any group of people, or any country, consider ourselves isolated; we are all in this together," he said afterward. There may have been nothing new in this idea, but after Gibson got back, he felt he had a firmer grip on the fact that this was one world than people who hadn't been in space. Similarly, Carr got concerned about air pollution in a way he never had been before, for when he looked out the wardroom window toward the earth's horizon, where he could see the atmosphere edge-on as a narrow arc, he was startled by how thin it was—it was, he said, like the skin on an apple. "Rather usually, most of the guys come back feeling a little more insignificant," he said after he had returned. "They see how big the earth is, and they think how short their stay is upon it, and what a small mark man has made on it. Most of the guys come back with an interest in ecology, for they see how much snow and desert there is, and how hard it is for the people who have to live there. You come back feeling a little more humanitarian. . . . People in our line of work—a very technical type of work—are inclined to move along with blinders on. You begin to get so involved with the details of what you're doing that I think you forget to look around you. And I think this mission is going to do me a lot of good in that I think it's going to increase my awareness of what else is going on."

In the end, even Mission Control saw that the astronauts had to have time to themselves. "At first, when they had begun asking for time off, I felt it was unwarranted," Hutchinson said afterward. "Those guys know how valuable their time up there is! Then I saw we'd done a bad thing by forcing them. I saw they needed time to think about what they were doing and to reestablish themselves. They were not asking for time to read beddy-bye stories!"

Hutchinson clearly had not been reading popular psychology books in *his* spare time, though he seemed to have gotten the right idea despite this lack. "We now see that time off is mandatory," he said. "A man has to get mental enjoyment out of something other than his work. We now feel that an astronaut's time off must be inviolate. And I think that before the mission was over, we reached a reasonable compromise. But we learned a marvelous lesson in how to manage people." It was a lesson, one flight surgeon said later, that almost anyone might have taught NASA earlier.

So in the latter half of their mission, Carr, Gibson, and Pogue got more time off to relax and do what they wanted —which usually meant looking out the window for their own enjoyment, not the earth resources scientists'. For these occasions, usually late in the evening, they sometimes saved their ice cream or butter cookies from dinner; Pogue wished he had peanuts and some candy bars. Similarly, of an evening, the skipper of Jules Verne's *Nautilus* would draw back the big curtains in the submarine's living room, and over refreshments, with which the *Nautilus* was better supplied than Skylab, he would point out the sights through a similar big round window. That window framed luminous fish, but when Skylab was on the dark side of the earth, the sky was full of glows, too, such as the airglow, the luminous layer on top of the atmosphere, or the noctilucent clouds, cloud decks the airglow illuminated from above. There were moments when the astronauts felt they were sailing over a milky, phosporescent sea. Pogue wished he had more film. "A lot of times we see things we'd like to take pictures of, but we're sort of reluctant to do it because it's sort of a 'gee whiz' picture, you know?" he said. "I'm not saying we should have our *own* film, but have film that's available, you know, so that you can just go ahead and take whatever you want." Sometimes, if they could see the glowing arcs and curtains of the aurora borealis, which were on the same level with them, the astronauts had a sense of motion, as though they were sailing through an Arctic sea of icebergs and rainbows. Sometimes the third crew could see the comet Kohoutek, a sweeping orange and yellow arc like the stroke of a brush that had been dipped in the sun; it reminded Gibson, whose imagination was more scientific than romantic, of the exhaust plume of a rocket lifting off from Cape Kennedy. It reminded Pogue of a dolphin. Above all, though, there were the stars, which were visible all the time, even in daylight, for they encroached to within thirty degrees of the sun. The astronauts never tired of looking at them, for above the atmosphere they had a hard, brilliant quality, and a variety of colors—reds, blues, greens—that the astronauts had never seen on earth. "When you're up here, you see the earth as one unit, you see the sun as a star, and you can see all the other stars out there, and you realize that the universe is quite big," Gibson said over the B channel. "The number of possible combinations that you could have out there which could create life, all this enters your mind and makes it seem very much more likely. I don't think this [idea] is any different from what people have thought down on the ground. It's just that being up here and being able to see the stars as you can, and look back at the earth, and see your own sun as a star, makes you much more conscious of the possibility." As Gibson was apt to stay up particularly late looking out the window, he was likely to be quite keyed up when he went to bed, and he sometimes had to read himself to sleep. He liked science fiction stories, one in particular that dealt with astronauts from another planet who, like himself, were surveying the earth from space.

Ten o'clock was normally bedtime. The astronauts' good nights to the ground, when they were in touch with it, could be quite lingering, as it was one night when Carr, Gibson, and Pogue, who had made up their differences with Mission Control, signed off with the team of flight controllers that was on duty, the Purple Team.

"Good night, Dick," Gibson said to the capsule communicator, Richard Truly, a fellow astronaut.

"Good night, Dick," Pogue said.

"Good night, Dick," Carr said. "Good night, Phil. Good night, Purple guys." Phil was Philip Schaffer, the flight director of the Purple Team.

"Good night, Phil," Pogue said.

"Good night, crew. Y'all drive careful, hear?" Schaffer said. It was unusual for a flight director to talk directly to a crew, but then the hour was late.

"We will, Phil," Gibson said. "Good night all to all the Purple people."

"Good night, Purple gang," said Pogue.

The CapCom said, "And Mrs. Calabash, wherever you are, say good night to Bill [Pogue]."

"Nighty-night, Bill," the flight director said in a falsetto.

"Well, y'all sleep tight, see you in the morning," the CapCom said.

"OK," said Pogue.

Before he turned in, Carr made a final tour of inspection to make sure that all was secure for the night; there was a checklist by his bed in case he forgot anything. He had to make sure that the flare alarm at the solar console was on. He set another alarm in case the ground had to wake them during the night, a third in case of fire, and a fourth in case the workshop was punctured by a meteor—in that event, there would be a drop in the space station's air pressure, and the astronauts would have to leap out of bed, find the hole, and, assuming it was small enough, plug it with a patch similar to those used for repairing tires before their atmosphere could escape. Although no noticeable meteorite pierced Skylab, and there were no fires, the astronauts frequently wished their bedrooms were closer to the command module, their only means of escape if anything went seriously wrong; and they urged that bedrooms on future space stations be not so far from the exit.

As they undressed, they pushed their golden shirts and pants through the slits in some soft black rubber knobs that moored them for the night; the knobs, though, were none too secure, so that their clothes floated away, and next morning they sometimes found their jackets, socks, and pants on the screens at the top of the workshop. During the second mission, Lousma had complained that even when his clothes remained fast to their moorings, they had tickled his face during the night; he had gotten so that he rolled them up into a ball and shoved them, safely out of the way, in back of his bunk. The astronauts' pockets still bulged with pens, penlights, scissors, and other items they had picked up during the day; ever earthbound, they missed a bureau top to lay them on.

Their beds, called sleeping restraints, were little more than light sleeping bags against the walls. Carr, Gibson, and Pogue leaped into them feet first, like genii returning to their bottles; the bags' necks were a little tight so that they had to wriggle a little to get through. Sometimes the astronauts were chilly in bed because the sleeping compartments were ventilated to the point of being drafty, lest in the absence of convection currents they suffocated in the carbon dioxide of their own exhalations while asleep. During the first mission, Conrad, who had been annoyed because the draft, which was from the floor, blew up his nose, and whose mechanical ingenuity was already much in evidence around the space station, had reversed his bunk in order to sleep upside down. Then, though, the draft blew up under his blanket which, without gravity to hold it down, had ballooned around him; as his body was unable to generate enough heat to warm the extra air, he was cold. Never at a loss for what he called a "fix," he had wanted to sew an extra blanket onto his bed, but he was unable to find a needle and thread aboard Skylab. He complained that all future space stations should have sewing kits.

Before they turned out their lights, the astronauts sometimes read a little in bed. Carr wrote in a diary each evening; he wished one of the cabinets in front of him folded out to make a desk. (Much later, when a reporter asked for a look at what he had written, he smiled broadly and said, "No way.") Then, before they went to sleep, they strapped themselves down so that they wouldn't float about inside their billowing blankets; they tucked their arms inside, and they even strapped their heads onto their pillows. Aside from the straps, there was no pressure on them, and they felt as comfortable as though they were lying on water beds. "I don't have [the restraints] on tight, just loose—loose enough to feel like I've got something there," Lousma had said to the B channel one night during the second mission. "Sometimes, I start to sleep on my stomach, or sometimes on my back. I can close my eyes and imagine myself in any position I want to, and position doesn't really matter much." The smoothest way of all to sleep, of course, was to float absolutely free, without any bedding or blankets or restraints whatsoever. Gibson tried this one night in the upper deck of the workshop, where he curled up in thin air, his body gradually assuming the neutral G crouch. He got caught in the air current, though, and was bumped awake every time he hit the screens in the dome.

Sometimes the astronauts had a hard time getting to sleep. Whenever the spacecraft passed from sunlight into darkness, or back again, the hull banged and popped loudly with the change in temperature, like a tin roof. Conrad had complained once that there seemed to be a whole new set of sounds in space. If Skylab's thruster jets fired during the night, they sounded like bursts of machine-gun fire. If anyone got up during the night, he invariably awoke the others. Garriott, the lightest sleeper, was most bothered by all the nocturnal disturbances. Kerwin and Gibson, though, managed to sleep more soundly than they did on earth. Kerwin didn't even remember having any dreams in space, though the engineers on the ground, who never left the astronauts alone even while they slept, say that he did.

As the third crew fired the thrusters to push its command module away from the space station, the little jets caused Skylab's solar panels to flap a little; the astronauts quickly turned the thrusters off and drifted several hundred feet away before firing them again. Pogue thought the enormous space station looked jewellike, because in vacuum, where no atmosphere got in the way, every little rivet stood out with a gemlike clarity. Skylab still contained a five-month supply of oxygen and a six-week supply of water, and its orbit was such that it should not plunge back into the earth's atmosphere for another eleven years. Pogue knew, though, that no one would ever again look out the wardroom window for hours at a time, as he had done, just watching the earth roll by with its variety of clouds and shifting color and light; this is what he would miss most. In the seat next to Pogue, Gibson, who had the job of delivering a final description of the space station, radioed, "It's been a good home," and the capsule communicator replied, "Yeah, it sounds like it; you guys occupied it long enough." This was the full extent to which anyone connected with Skylab would reveal his feelings upon the abandoning of a craft in which nine men had spent a total of 171 days, circled the world 2,476 times, and traveled 70,500,000 miles—many times farther than the total distance traveled by all the Apollo astronauts who went to the moon, almost as far as from the earth to the sun. Pogue watched the space station as long as he could; when it was so small that he could cover it with his thumb at arm's length, Skylab apparently was listing on its side in relation to the earth, the telescope tower still pointing at the sun but parallel to the ground, its solar panels and its remaining wing vertical to the earth. Then Pogue's attention was diverted, and he never saw the space station again.

Most of the astronauts left Skylab with mixed feelings. Conrad had said at the end of the first mission, "We had schooled ourselves for twenty-eight days, so we were ready to go home; but it had been an extremely comfortable home while we'd been there, and we sort of hated to leave it, to get back to one G." Although all the Skylab astronauts insisted they could have stayed there indefinitely, most of them were ready to come home when their tour was up. "I never got tired of looking out the window; I never got tired of zero gravity," Conrad had said. "I knew I'd never get back up there, but I was ready to come home. . . ." He said that if on the last day the ground had called up and said, "Surprise, you're going to go for another sixty days," he wasn't quite sure how he would have felt. Most of the astronauts thought that Skylab was a nice place to visit, but they didn't want to live there.

The return to earth, and subsequent readaptation, of the first crew was the most difficult, contrary to all the flight surgeons' expectations. The moment they worried about the most was the actual reentry, when the earth's atmosphere would be slowing the command module, for it was then that Conrad, Kerwin, and Weitz would have to begin to pay the penalty for their adaptation to weightlessness. Deceleration forces were virtually indistinguishable from gravity, and the flight surgeons, who didn't want to place too great a load on the astronauts, saw to it that they made two smaller rocket burns instead of one big one, so that the gravitational forces on them would be reduced when they slowed the command module to take it out of orbit. Even though the first burn provided a sensation only three-quarters of the earth's gravity, Kerwin almost fainted—he began to gray out and get tunnel vision, an inability to see to the side as a result of the unaccustomed rush of blood to his feet. Part of the trouble was that Kerwin had never completely inflated a pair of rubber pants he, and the other astronauts, were wearing, which would have exerted a restraining pressure on his lower legs to prevent any pooling of blood there. The two others had not been so forgetful. Doctors frequently fail to pay enough attention to their own health, and Kerwin proved no exception in this regard. Weitz experienced some slight tunnel vision, too, but Conrad did not. No other astronauts grayed out during entry. Kerwin, who up until then had felt what he called "bouncy" in space, said later that the episode "surprised the tar out of me." Kerwin proved to be in about the worst condition of any of the returning Skylab astronauts; Conrad was in about as good condition as any; and Weitz was in between. The dramatic differences among the three men on the first crew happened to coincide with the amount of time they had spent exercising on the ergometer in space—again with Kerwin, his protestations of bounciness to the contrary, having put in the least time at it. Kerwin's typically doctorish failure to look after himself more carefully was helpful scientifically, if not to himself, for it was the discrepancies among the three men that fortified the flight surgeons' belief that exercise was essential to astronauts' well-being on long missions, and caused them to recommend that the two later crews spend more time on the ergometer.

Aboard the U.S.S. *Ticonderoga*, the aircraft carrier that picked the men up, Conrad, Kerwin, and Weitz had to walk from the command module down a red carpet to a medical laboratory about thirty yards away. As soon as they stood up, they felt faint because of the blood rushing to their feet. When they took their first step, they felt unusually heavy, as though their arms and legs were made of lead. Conrad wasn't certain Kerwin could make it. Conrad himself felt better than he had after his fourteen-day trip to the moon; he walked down the carpet in a fairly straight line. Weitz wobbled a little; he was walking in a deliberately broad-based way, his legs far apart, to keep his balance. Kerwin, though, couldn't walk in a straight line at all; his gait, one of the flight surgeons who accompanied him thought, was ataxic—he was unable to put one foot in front of the other, but rather he lurched tipsily from side to side. Not only were the astronauts unaccustomed to walking, but many of the muscles they used for that purpose, in particular their calf muscles, were unexercised, the ergometer notwithstanding. Their legs didn't respond properly. Many months later an astronaut returning after the second mission said, "Somehow the little sensors in the legs that want to keep you upright and respond to lateral motions just hadn't been exercised in a long time, and they had not gotten the word that we were back on earth yet; and so if you went into a swerve, you probably stayed in a swerve and ricocheted off the wall, or you had somebody catch you." They felt as out of control as they had when they had first arrived in Skylab. Another reason for their stumbling was that, to varying degrees, they all had vertigo, because for the first time in several weeks gravity was influencing the fluid of their inner ears. Conrad, who sometimes thought of his body in the same terms as a spaceship, said he felt as though "my gyros are tumbling"—what happens when a gyroscopic guidance system loses its bearings. Again, the difference was similar to what they had experienced the first few days they had been in space when the liquid balance of their bodies, and of their inner ears, had been changing. As had happened in space, these flashes of vertigo were over quickly, reflecting the rapid return of the astronauts' liquid balance to normal. These flashes lasted about two hours with Conrad, eight hours with Weitz, and about two days with Kerwin. In spite of these troubles, though, all the Skylab astronauts were in far better condition after their return than the Soviet astronauts who, after just eighteen days in space, had had to be removed from their craft on stretchers. Kerwin thought that the difference was probably due to the large size of Skylab, which allowed the American astronauts to move around and even exercise; he said later that if *he* had been as constricted as the Russians aboard their craft, he might well have had to be carried out on a stretcher, too. NASA, indeed, was prepared for this eventuality.

Inside the medical laboratory, the flight surgeons made the astronauts lie down. They took their pulse and discovered that their hearts, which had averaged about twenty beats lower than normal while they had been in space, were now about ten beats higher than normal; their systems had been operating at a slow speed for so long that it required a little extra effort to maintain themselves on earth, where pumping blood was harder anyway. When the astronauts stood up, the change was even more pronounced. Because of the weakened veins and arteries in their lower bodies, more blood pooled at their feet, and consequently their hearts had to pump even harder to get a sufficient supply of blood to their heads. The doctors thought that all three astronauts could black out; once again, Kerwin came closest to doing so. They put more air into his rubber pants. Kerwin shortly threw up as the result of having ill-advisedly chug-a-lugged a cherry soda in the command module; he sat in a corner looking miserable. Conrad and Weitz, though, were soon well enough to try the bicycle ergometer; Conrad's performance was off from his preflight level by ten percent and Weitz's by fifteen. When Kerwin tried the ergometer the next day, though, his performance was off by twenty-five percent.

Generally, all the trends reversed themselves in about the same order as they had occurred in space. The fluids that had migrated to the astronauts' upper bodies returned to their lower bodies before the end of their first day back. With the fluids no longer pooling about their hearts and giving them false feelings of satiety, the astronauts were thirstier; they drank more water, and they retained it. As a result, those other trends that had de-

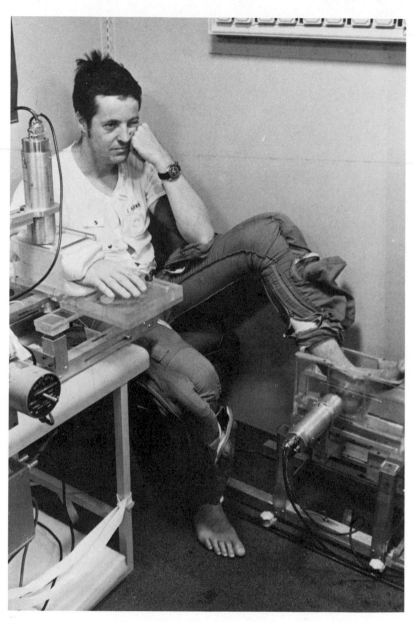

Back on earth Kerwin sits in a corner looking miserable.

pended on the loss of body fluids, such as the astronauts' electrolyte balance, returned to normal quickly, too. By the end of their second day back, they had regained the liquids they had lost in space, which had accounted for about half of their total weight loss—between four and eight pounds; the other half, which reflected the loss of muscle tissue, took another two or three weeks to regain. Even so, their strength came back rapidly: at first, they had felt the weight of the earth's gravity upon them at about three times what it ought to have been—that is, they themselves had felt three times too heavy. But by the second day, they felt they were only twice too heavy, and by the third day they felt they weighed the right amount. As the astronauts' strength returned, they suffered from sore muscles. The ones that ached the most were, of course, the ones they had used least in space—in addition to their calves, their lower backs hurt, and so did their neck muscles, which were once again supporting the unaccustomed weight of their heads. An old back ailment of Conrad's recurred, because muscles that had compensated for the injury had lost their strength. Six days after they got back, though, the first crew was playing paddle ball, and Conrad at least, despite his back trouble, was close to performing on the ergometer, and also in the LBNP, as well as he had before he had gone into space. The astronauts' pulses were still about five beats higher than they had been before the mission, and they wouldn't return to normal for another week, which was when their hearts presumably regained their preflight strengths.

The second and third crews were in much better condition when they returned, and their recovery was correspondingly faster than the first crew's. The second crew, some flight surgeons estimated, was in about twenty percent better condition than the first crew, though it had been in space twice as long; and the third crew, which had been up half again as long as the second, was generally considered to be in the best condition of all. For example, it took the first crew, held back principally by Kerwin but also by Weitz, sixteen days to return to its preflight level on the ergometer, but the second crew was back to normal on the fifth day, and the third crew on the fourth. Other trends followed the same pattern. There was almost nothing about the third crew that would have led an outside observer to suspect that these men had spent three times as long in space as the first crew; in fact, an outsider would have had ample reason to conclude that the *first* crew had been there the longest. The second and third crews, of course, had done the most exercise; indeed, the third crewmen, who had done the most exercise of all, recovered so quickly that on their fifth day back Dr. Hordinsky, the crew surgeon for the third mission, said that a doctor unfamiliar with them would have had a hard time knowing that they had been in space at all—except for some small clues, such as anemia (one of the last trends to return to normal was the astronauts' blood count). Even in this respect, though, the third crewmen were better off than the others, for they had already begun to produce new red blood cells in space.

Indeed, the medical results of Skylab had not remotely lived up to the predictions, which had been that the crew that was up the longest would be the worst off when it returned. It was puzzling. Although the flight surgeons were convinced that exercise was an important factor in a crew's health, as the results seemed to bear out, they had already begun to suspect, from such things as the blood counts of the returning astronauts and their improvement on the LBNP while they were still in space (which did not have much to do with exercise), that physical fitness was not the whole story. It occurred afterward to Dr. Hordinsky and some of his associates that by concentrating so hard on the amounts of daily exercise the three crews had respectively done, they might have failed to attach any meaning to a fact that had been jumping out at them all the time: *the healthiest crews also happened to be the ones that had been in space the longest.* The astronauts' conditions, of course, had deteriorated until they had leveled off at about the fortieth day, but after that, Dr. Hordinsky now noticed, it looked as though the longer an astronaut had been in space, in many respects, the better he was when he got back. "You throw someone in a new environment, and he's apt to have a tough time at first; but if he survives, he will tolerate it, and then begin to improve," Dr. Hordinsky said. "It could be that the crews that were up the longest had the longest time to recover! It might take between three and six weeks for a man to really adapt to space." The flight surgeons were beginning to think that the twenty-eighth day in weightlessness, which was when the first crew had been brought back to earth, was also the time when a crew was at its lowest ebb. If the second and third crews had been brought back then, they felt, those astronauts would have been as badly off, and conversely, if the first crew had come back later, its condition would have been better.

Although the flight surgeons, and everyone else around NASA, were overjoyed with this unexpected diagnosis, the doctors were not about to clear a man for an open-ended stay in space. The most they would consider, when Skylab was over, was a flight of six or eight months, about twice as long as the third Skylab mission. (There will not be such a flight in the foreseeable future, for NASA has no manned flights planned beyond the space shuttle, which can't stay in space more than thirty days —an awkward duration, if Skylab is any guide.) Doubling was the same cautious multiple by which the flight surgeons had always permitted the lengths of missions to be increased, once a given term in space had been shown to be safe: the Mercury flights had lasted no more than seven days; the Gemini and Apollo flights had lasted less than fourteen days; and the first Skylab flight lasted twenty-eight days, with the second originally scheduled for fifty-six, though it was later extended to fifty-nine. There were still too many unknowns to allow a greater increment. Perhaps some of the trends would start up again. Perhaps there were other trends, particularly of a cellular nature, that had not been discovered yet. Perhaps the loss of the sense of balance might reach a point where it was irreparable—as it was, some astronauts inexplicably retained their ability to rotate in the revolving chair with immunity for a month after their return. Nor were the flight surgeons altogether happy with what they did know, such as the large amount of red cells some of the astronauts had lost, which in Garriott's case had reached twenty percent. And they were edgy about the fact that in the case of one member of the third crew, fifty percent of his red cells had become thin and attenuated—something that didn't seem to harm him, but that the flight surgeons wished to know more about.

The chief worry, though, was one trend that had not stabilized in space at all: the slow, steady loss of calcium from the astronauts' bones, which had gotten progressively worse the longer they had been weightless. This was the only area in which the two later crews were worse off than the first, and it was the trend that took the longest to right itself after their return. On the average, the first crewmen lost only a small fraction of a percent of their bone calcium, but Garriott on the second crew and Pogue on the third each lost seven percent. This is not in itself a dangerous amount, but men who have been in bed, a condition in many respects similar to weightlessness, for nine months have lost as much as forty percent of their bone calcium, which *is* a dangerous amount, and Garriott and Pogue had progressed well along the same curve. The flight surgeons' official report on the matter read: "Capable musculoskeletal function will be threatened during prolonged space flights lasting one and a half years to three years, unless protective measures can be developed." These might involve drugs that would retard the calcium loss, or the selection of astronauts who were less susceptible to it; but the surest way of controlling the problem on long missions would be to have gravity in the spacecraft. After Skylab, there were people at the Space Center who thought this would be the best solution to a number of problems; Pogue, who was still of that opinion, said, "In the shape I was in when I landed, I could not have walked out onto the surface of, say, Mars and done good work." Indeed, in its recommendation that the nation look into building a huge space station in the future, the conference which NASA held at its Ames Research Center a year after Skylab was over reverted to the idea that the space station be shaped like a wheel that would rotate at one revolution per minute, fast enough so that the centrifugal force at its rim—where the living accommodations, schools, shops, and factories for its ten thousand space colonists would be located—would be the equivalent of normal gravity on earth. (The "gravity" would diminish toward the hub, where a man would be weightless.) According to the conference report, the space station would cost about a hundred billion dollars, be about a half a mile in diameter, weigh (if it were on the ground) about five hundred thousand tons, and have approximately the mass of the largest supertanker. When its day comes, if it ever does, Skylab will seem as archaic and fragile as the *Mayflower*. Regardless, the astronauts aboard Skylab showed that man is a more adaptable creature, and space a more suitable home for him, than anyone had previously expected.

After the astronauts' bone calcium had returned to normal, there was no way a flight surgeon could tell by any clinical test that any of them had ever been in space. Yet as much as six or eight weeks after their return, the astronauts' wives reported, they stumbled at night in the dark, evidently requiring a visual clue to the room's vertical, even though their sense of balance had completely returned to normal. And for a long time afterward some of the astronauts kept trying to float things around them as they had done in the space station. One morning when he was shaving, Lousma tried to leave his can of shaving cream hovering in midair. It crashed to the floor.

"It's hard now to put myself back into Skylab," Lousma said several months after that. "Sometimes I go over to the Skylab trainer, or sometimes I watch the movies we made, but I cannot re-create what it was like. It's almost impossible to recapture now. I never dream about the Skylab. But I often wish I could go that way again."